COMPUTER MODELS IN ENVIRONMENTAL PLANNING

COMPUTER MODELS IN ENVIRONMENTAL PLANNING

Steven I. Gordon

Department of City and Regional Planning
The Ohio State University
Columbus, Ohio

VNR VAN NOSTRAND REINHOLD COMPANY
——————— New York ———————

Library of Congress Catalog Card Number: 84-13031
ISBN-13: 978-1-4684-6592-1

Published by Van Nostrand Reinhold Company Inc.
135 West 50th Street
New York, New York 10020

Van Nostrand Reinhold Company Limited
Molly Millars Lane
Wokingham, Berkshire RG11 2PY, England

Van Nostrand Reinhold
480 Latrobe Street
Melbourne, Victoria 3000, Australia

Macmillan of Canada
Division of Gage Publishing Limited
164 Commander Boulevard
Agincourt, Ontario M1S, 3C7, Canada

15 14 13 12 11 10 9 8 7 6 5 4 3 2 1

Library of Congress Cataloging in Publication Data

Gordon, Steven I.
   Computer models in environmental planning.

   Includes index.
   1. Environmental policy—Mathematical models.
   2. Environmental policy—Data processing. I. Title.
   HC79.E5G655   1984   363.7′0525′02854   84-13031
   ISBN-13: 978-1-4684-6592-1   e-ISBN-13: 978-1-4684-6590-7
   DOI: 10.1007/978-1-4684-6590-7

To my parents, Jack and Rose Gordon; and to my family, Gaybrielle, Ben, and Tim—for their love and support.

Preface

The purpose behind *Computer Models in Environmental Planning* is to provide a practical and applied guide to the use of these models in environmental planning and environmental impact analysis. Models concerning water quality, air quality, stormwater runoff, land capability evaluation/land information systems, and hazardous waste disposal are reviewed and critiqued. I have tried to emphasize the practical problems with data, computer capabilities, and other analytical questions that must be faced by the practitioner attempting to use these models. Thus, I do not delve too deeply into the theoretical underpinnings of the models, referring the reader instead to specialized references in this area. For each environmental area, I review the major models and methods, comparing their assumptions, ease of use, and other characteristics. Practical examples illustrate the benefits and problems of using each model.

Computer models are increasingly being used by planning and engineering professionals for locating and planning public works, and industrial, commercial, and residential projects, while evaluating their environmental impacts. The requirements of the National Environmental Policy Act and related state laws as well as separate state and federal laws concerning air and water quality, stormwater runoff, land use, and hazardous waste disposal have made the use of these methods mandatory in many circumstances. Yet, explanations of both the benefits and problems associated with supposedly easy-to-use computer versions of these models and methods remain, at best, difficult to retrieve and, at worst, incomplete.

A number of federal and state agencies have invested rather heavily in the development of computer models. In some instances, one can make the case that the investment has been wasted. The documentation may be impossible to decipher, if it exists at all, or there is no correlation between the documentation and the computer code. One does not wish to waste time trying to learn that this is the case. In many other cases, these problems do not occur. In fact, there are many computer models that are well documented, readily transportable to other systems, and of great use in answering the required environmental questions under state and federal laws and regulations. Unfortunately, the information on which models are available, how they can be acquired, and which are worth the investment in time, effort, and other resources is scattered among a wide array of government documents. This makes it quite difficult for those not intimately connected with the development of the models to acquire and use them.

To this end, I have tried to assemble in one place a comprehensive review of the major modelling efforts in the environmental planning arena. I hope it will help practitioners to incorporate such models into their work, and that it will help students of environmental modelling obtain a useful overview of the practical aspects of simulation modelling and impact analysis.

STEVEN I. GORDON

Acknowledgments

Several people and organizations deserve recognition for their help in the production of this manuscript. First, I must thank Steven Arend, Scott Eberhart, Robert Johnson, and Bruce Riffle for their hard work in producing many of the illustrations used in the book. Second, Ellen Wallace and Madonna Alessandro must be credited with assisting with typing of much of the tabular and mathematical material in the original manuscript. Two organizations also deserve some recognition. The Ohio State University Task Force on Learning provided the original small grant that helped me to obtain and incorporate into my teaching and research efforts many of the computer models reviewed in this book. Finally, I must thank the W. K. Kellogg Foundation, who, through a National Fellowship, helped me to acquire some additional expertise in areas new to me and the IBM Personal Computer on which I composed most of the manuscript.

Any ideas and opinions are, of course, my own and should not be attributed to these individuals or organizations.

Contents

COMPUTER MODELS IN ENVIRONMENTAL PLANNING

1
Models in Environmental Planning

INTRODUCTION

An engineer and planner calls a meeting to discuss the potential for a water supply and flood control project in upstate New York. After outlining the nature of the project, someone from the audience asks what the environmental impacts of the projects might be. The engineer/planner responds, "Don't worry, we'll count all the birds and the squirrels and the bees." What this individual is referring to is the requirement to undertake an environmental impact assessment relating to the project. Unfortunately, this individual has a misconception of what this process entails and how to go about it. Fewer individuals, however, currently misunderstand the legal requirements associated with the environmental laws because of the last decade of experience with them. In addition, this experience has produced more professionals who plan projects without major, negative environmental consequences.

The requirements for such assessments have already been alluded to—State and Federal environmental regulations. These include the National Environmental Policy Act, the Clean Water Act and amendments, the Clean Air Act and amendments, the Resource Recovery and Conservation Act, and several other pieces of legislation. The legal requirements embodied in such laws have been extensively explicated in various books and manuals. The major question which must be faced by the analyst approaching the requirements in good faith is what methods are best for answering the critical questions associated with protecting the natural environment.

Cynics among us might contend that there are no good methods and, worse yet, that the agencies involved in carrying out these laws have not discovered that fact. In reality, a number of approaches to analyzing environmental impacts have come to be accepted in spite of their faults as a practical way of meeting the intent of the environmental laws. These laws will not be reviewed in detail in this book.

However, they will be referred to in relation to the types of analyses that might be carried out in the process of planning for major new facilities. In particular, these laws provide specific limits on the nature, amount, and location of waste discharges into the environment that are, in turn, related to the changes in quality of the ambient environment—the land, watershed, or airshed into which the wastes are discharged. Permission to operate many types of new facilities hinges upon the demonstration that the environmental impacts lie within the acceptable ranges of environmental quality provided for by these regulations.

Regulations are important not only at the State and Federal level but at the local level as well. Local land use controls frequently require the developer to meet local performance standards with regard to air and water quality, noise, and storm drainage. Special ordinances, such as storm drainage ordinances, require site designs to include methods for controlling the increased stormwater on site and thus require modelling of the local storm drainage system. Regardless of the status of local ordinances, local officials often need to be reassured that the benefits of a new development will not lead to unanticipated local costs for increases in public facilities; thus they wish to see an analysis of the impacts of development on various local facilities.

At various geographic levels then, both regulatory and political considerations require the analysis of the environmental impacts of proposed public and private development actions. Because of the complexity of both the natural environment and the design alternatives available, consideration and analysis of all but the most simplistic solutions is very difficult without the aid of a computer. Although many public and private resources have been expended on the development of computer environmental models, the availability of such models is limited by the lack of a widely known and central distribution point, by poor documentation, and by a lack of information on the applicability of various pieces of software to particular situations.

The purpose of this book is to introduce and review some of the major model choices available to the planning and engineering professional. A number of models will be reviewed in five major areas: water quality, stormwater runoff, air quality, land capability analysis, and hazardous waste disposal. In each case, available software will be presented and reviewed with regard to the ease or difficulty of acquisition, costs for adaptation and use, applicability to various situations, and technical requirements for use. The intention is to provide a guidebook that will help practitioners take advantage of computer technology in environmental analysis and avoid as well some of the pitfalls associated with using software developed for special applications.

MODELS AND DECISION MAKING

Physical vs. Mathematical Modelling

The complexity of both human and natural systems has led to the use of models. In this sense, a model is just a simple representation of the real world. Models can be of many forms. Physical models provide a physical analogy to the operation of a system. One of the best known of these is the Army Corps of Engineers' model of the Mississippi River in Vicksburg, Mississippi. This model allows the simulation of major flood events, showing physically where the flood waters flow and the nature of the flood damages that might occur.

In most instances, physical representation of the system being modelled is not possible. In this case, a mathematical model is generally used. There are many classes of such mathematical models. (See Bertalanffy, 1968 for a review.) The underlying rationale behind all such models is to allow for the analysis of the operation of some system in order to evaluate system performance. The researcher then uses the model to determine whether or not the system will function at some desired level. The model will, of necessity, limit the number of variables that are simulated, trading some realism for ease of use. Some acceptable level of accuracy should be maintained if the model is to be acceptable for the application at hand. Thus, a water quality model greatly oversimplifies the operation or performance of a stream system in order to study the reaction of certain critical variables (e.g. dissolved oxygen) to alternative management decisions.

Advantage of Computer Modelling

The management decisions made by planners include those related to land use, the provision of public services such as water and sewerage, and the impacts of new facilities, both public and private, on the natural environment. In making decisions related to these facilities, planners have traditionally not utilized computer models to a large extent. Instead, relatively simple calculations have been combined with personal judgement and experience to arrive at a "gut" feeling about what the "best" decision may be.

The problem with this approach is that it limits the complexity of the decision making framework since most people cannot intuitively analyze more than a few variables at one time. As a corollary, this approach often fails when one deals with complex systems for which intuitive decisions are overly simplistic. For example, one may choose the "best" site for a new power plant based on the relatively obvious criteria of inexpensive land, proximity to adequate water supplies, and proximity to the existing electricity market area, only to find out that the air and water quality impacts of the plant at that site will be unacceptable to local and Federal authorities.

Once resources have been allocated to the design of a facility for that site, a shift to a more favorable location from the environmental perspective may be difficult. Even if a decision to move the plant is made, valuable resources will have been wasted because a more careful analysis of locations had not been made. Computer modelling offers a relatively inexpensive and effective way to incorporate a larger number of variables into the decision process and thus avoid serious resource allocation errors.

Decisions which must be made with regard to environmental impacts tend to involve a large number of criteria because of the complexity of environmental systems. Thus, regardless of the scale at which decisions are going to be made, regional or local, environmental systems are too complex to allow for intuitive decision making. They are large scale systems with a number of characteristics:

1. A large number of decision variables, exogenous variables, and state variables;
2. A large number of components (subsystems);
3. A complex and often nonlinear functional input/output relationship;
4. Risks and uncertainties;
5. A hierarchical organizational structure;

6. Multiple, noncommensurable, competing, and often conflicting objectives;
7. Multiple decision makers; . . . (Haimes, 1982, p. 2).

Unfortunately, the converse argument relative to the use of models can also be made—that models are inaccurate representations of reality which distort decisions involving valuable planning resources. In the past, this has often been true for various reasons. Frequently, models have been inadequately documented to allow for easy adoption by others. When such adoption has occurred, major errors have been made arising from inappropriate model use. More often, resources have been expended on the formulation of sophisticated models only with the result of the models' abandonment in the long run because the decision makers involved could not understand how to use them, or the models did not meet their needs. Examples of these situations are legion (see Fisch and Gordon, 1980; Andrews, 1978; Haimes, 1982).

Avoidance of Pitfalls

These continuing problems aside, models can still be valuable tools for the analyst attempting to consider more than simplistic approaches to making planning decisions impacting the environment. However, in order to avoid the principal pitfalls of model applications, even if previously successful, a number of criteria should be used to evaluate any model taken under consideration. First, we can define two major, overriding model requirements:

1. A model is useful to planners only if it simulates the impacts of realistic decisions.
2. Any model must be validated as an accurate representation of the real world within some defined limits of reliability before it is used to make decisions.

The first requirement seems logically obvious but still has not prevented some analysts from developing and attempting to use some marginally important models. For example, one government sponsored study used a linear programming approach to analyze the optimum distribution of hikers on a wilderness trail. The objective was to schedule the hikers so as to minimize the possibility of their meeting others on the trail, and thereby "maximize" the wilderness experience. No doubt this is a noble objective. Yet, when one thinks about it, the effort required to develop, test, and run the model seems a bit overblown in light of the fact that creating a management system to carry out such an optimization scheme is not realistically feasible, even if possible at all. Another less esoteric example is a demographic model which projects population for five year periods by sex and five year age groups, but does so for multi-county regions of the U.S. which have no administrative or geographic meaning.

The second objective also seems obvious but again is not necessarily adhered to by model builders and users. Most typically, model results are compared graphically with known values. If the lines appear to be more or less parallel, the model is said to be validated. Figure 1-1 illustrates this approach for a water quality model. The model here is said to perform satisfactorily since the lines on the graph have approximately the same form. The problem with validations of this nature is that they are entirely qualitative while the model results being applied for policy decisions is quantitative, leading to recom-

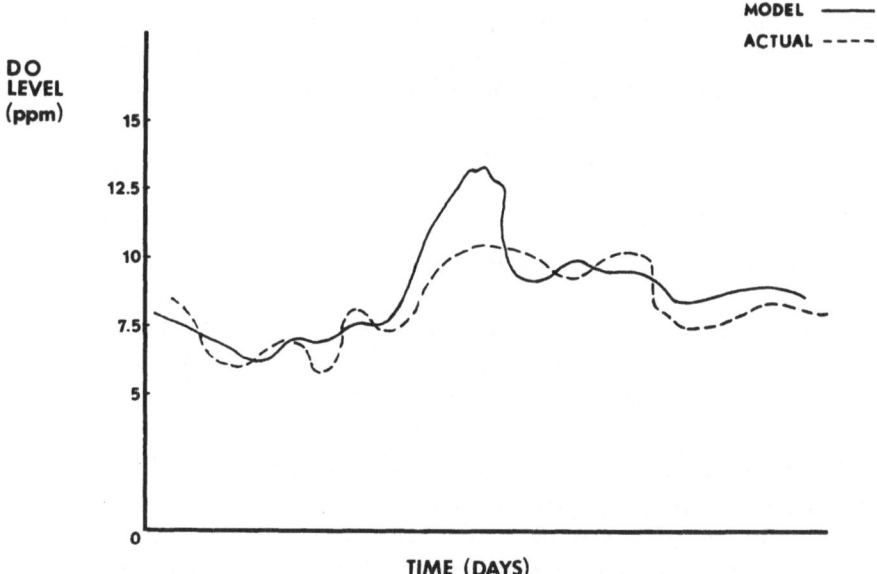

Figure 1-1. Water quality model vs. known values.

mendations that involve new facilities with definite capacities. If one wants sewage treatment to meet minimum water quality standards, errors associated with qualitative validation may not be acceptable. Without a more definite idea of the nature and quantity of model errors, public officials and professionals using the models will have little confidence that their decisions arising from the model results will be correct ones.

Validating Models

A number of quantitative approaches to the validation of models are available. Thomann (1982) makes a good review of these using some water quality examples. The methods include regression, relative error, comparison of means, and the root mean square error. These measures will not be reviewed in detail here but several will be used to evaluate the efficacy of various models discussed in subsequent chapters. All of these measures of model accuracy are quantitataive in nature, allowing the analyst some feel for the overall accuracy of model results and providing some confidence that the decisions being recommended are within the realm of acceptable error levels.

What these first two model requirements show is that every model must be carefully evaluated before it is used in order to ensure that the proper model is chosen for the task at hand. The steps required in such an analysis are illustrated in Figure 1-2 and would include the following:

1. Delineate those factors known to influence the system being modelled and in what ways they vary;
2. List those system components which are included in the model;
3. Explicate the implicit and explicit model assumptions based on the model documentation and a comparison of (1) and (2).
4. Given a preliminary evaluation of model capabilities and an understanding of the planning decisions which must be made, determine if the model will be useful for the purposes at hand. If not, go no further with it.
5. Determine the requirements of the model in terms of input data, data manipulation, type of computer equipment, computer time. If these exceed available resources or capabilities, go no further.
6. Make the model operational on the analyser's computer system.

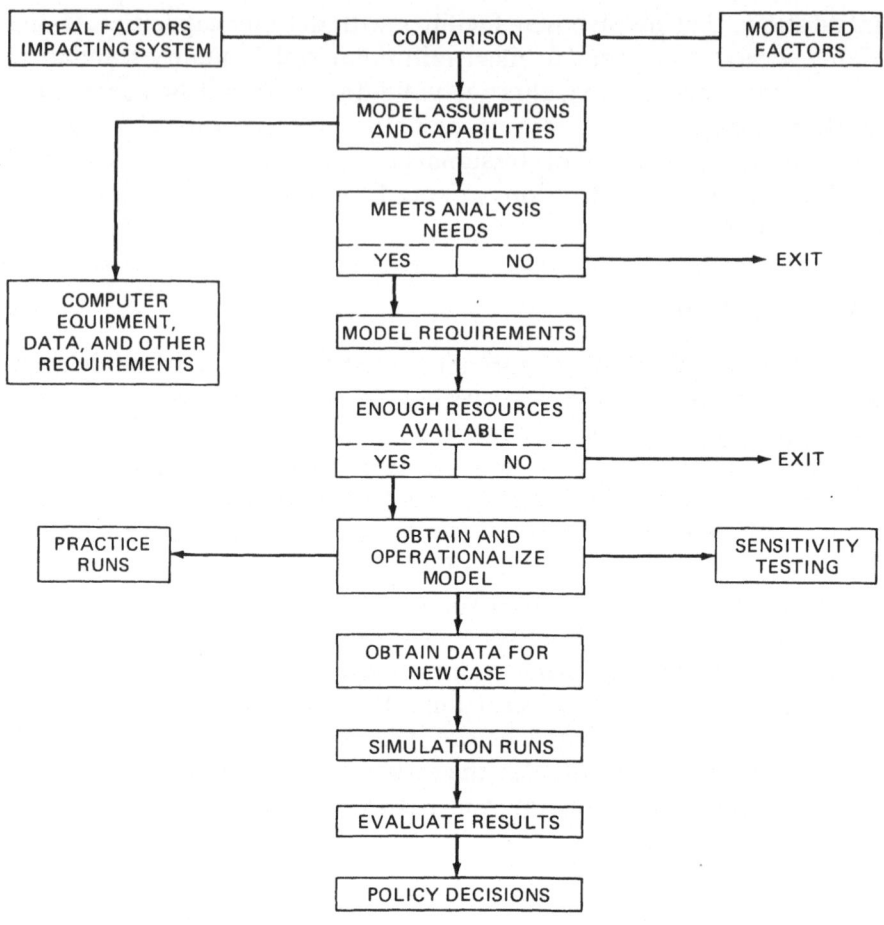

Figure 1-2. Model evaluation process.

Make certain using an example data set, that all options work properly and that use of the model from a purely mechanistic standpoint is fully understood.

7. Test the sensitivity of the model to variations in the basic parameters. Does the model react like the real world is known to behave? Are the upper limits of the physical world incorporated into the model, or does the model exceed known limits beyond some point? If the model is insensitive to parameter variation, is this an accurate representation of reality or only a mathematical accident?

8. Given that the model appears to be "acceptable," use it to simulate the decisions at hand.
9. Evaluate the potential results of various decisions bearing in mind all the assumptions and sensitivities of the model.
10. Choose the alternatives which "best" fit the priorities and goals associated with the decisions being made.

This list is certainly not exhaustive of the types of model evaluations that could be undertaken. It does provide, however, a minimum list of tasks that should be undertaken in order to avoid some of the major errors and pitfalls associated with applications of mathematical models.

A review of the list will make apparent that many tasks should be undertaken well before a major modeling and policy effort gets underway. The model cannot compensate for the ignorance of the analyst with regard to system behavior. Thus, the first tasks are aimed at forcing the model user to understand fully, to the limits of the "state-of-the-art," the nature of the system being modeled. These steps also provide an indirect check on the honesty of model documentation. Are the assumptions of the model explicitly stated in the documentation or are they hidden extras? In some ways, the person hawking computer models should be viewed in the same way as a used car salesman in order to maintain a healthy understanding of what is being purchased.

Checking Resources for Model Use

Even after the model has been found to be a realistic approximation of the real world according to the documentation and useful for the decisions at hand, one must make a realistic determination of the resources that will be required to implement the model on the local computer. Even models that can be implemented with free software cannot be used without cost. The computer on which the model is operational may have a particular compiler that is not available at the analyser's particular agency or firm. This implies the availability of some time to edit the program to fit the locally available compiler or to lease time on a compatible computer elsewhere. If the program comes on computer tape, there may be problems reading the tape, requiring again the use of outside services for tape translation. Every model of course requires an adequate set of input data. If no data are available for an area, at what cost can they be collected (if at all). How long and at what cost can available data be translated or transferred into the format required by the model? How flexible is the model with regard to data requirements and formats? There are a large number of technical and cost questions which must be satisfactorily answered before one proceeds with model use. Appendix A looks at some of these questions in more detail while subsequent chapters attempt to address them with regard to the particular models discussed.

Getting the Model to Run: Sensitivity Testing

Only at this stage—resolution of technical and cost questions—can the model be obtained for actual use. Now, some additional initial model tests can be run. If the documentation does not provide an indication of the sensitivity of the model to parameter variation, this will need to be the first task. The focus here is on several types of parameters: parameters assumed to be constant by the model or having a finite and relatively small number of values; the incorporation of the physical boundaries of the real world into the model; and the relationship between parameter insensitivity and real world conditions.

The first item is what is most frequently referred to as *sensitivity testing*. Here, various parameters in the model are forced to take on known extreme values to see how they impact the model results. If very little change occurs in model output, the model is said to be insensitive to variations in that parameter. This effect is often used as justification for assuming that a particular parameter can be held constant with little or no impact on model accuracy, which, however, may not be the case unless the two other questions are answered satisfactorily.

Since we are talking about environmental models, there are physical limits to the values that can be assumed by certain parameters. Some are defined by the limits of the laboratory tests used to measure them. This is the case for various chemical compounds. Others are limited by chemical relationships in the real world. One cannot have a negative amount of any physical quantity while certain variables have natural upper limits. For dissolved oxygen in water for example, this is normally 14.6 parts per million. Surprisingly, some models do not incorporate these physical limits as constraints.

Although it is not always possible to check, one must also try to confirm that the real world behavior of the so-called insensitive parameters are indeed what goes on in reality. Are there experimental data which confirm that this type of insensitivity does exist? If no such data are available, will the policy questions at hand provide any conditions wherein an assumption of insensitivity might result in erroneous conclusions? As we will see, such assumptions can become problematic in certain situations.

Undertaking any sensitivity testing must of course be preceded by making the model operational on the local computer. Sometimes this can be problematic depending on the computer resource requirements of the model. Several types of both hardware and software problems might occur, including insufficient memory, incompatible input/output devices (tape drives, plotters, etc.), or incompatible versions of computer code (compilers). Some of these problems may be avoided by an understanding of what the local limitations are before one orders a copy of a program available in multiple formats, while other problems may be completely unavoidable. Appendix A looks at some of these problems vis-à-vis the models reviewed in this book.

Adding the Data Base

The next and equally important step after getting the model running is preparing the data for the policy analysis question at hand. For some models this is quite straightforward and involves only the application of local knowledge of existing or proposed facility design, some easily obtainable published data, or some relatively easy field work. In other cases, a much more extensive level of effort may be required. The required inputs may involve the use of published data that are not centrally available or, if available at all, are not "clean." This means that data are missing, in the wrong units, not in computer readable form, or not reliable. In such situations, the production of an acceptable data base will be very labor intensive and thus expensive to accomplish. In even worse circumstances, the model may demand so much data as to require a major data gathering effort over an extended period of time. Thus, data requirements and availability must be carefully considered in selecting and implementing a computer model.

Following all of these trials and tribulations, the model can finally be used to simulate alternative decisions and arrive at a set of recommended actions, impact analyses, or other related analyses that will be used in the decision making process. Here, all of the previous work will be important in guiding the model user with regard to accuracy, reliability, and quirks of the model, and in thus avoiding inappropriate or erroneous recommendations.

One may not need to undertake all of these actions for each model being used since many have been appropriately tested and documented under a number of different situations. In fact, many of the models reviewed in the remainder of this book have been tested sufficiently to allow more immediate adoption and use. However, the analyst must be aware of potential model problems in order to make realistic recommendations based on model results. Thus, many of these questions are reviewed with regard to each of the models discussed in the following chapters.

ORGANIZATION OF THE BOOK

In each of the five chapters that follow, a set of computer models will be reviewed in detail. Chapter 2 will focus on several water quality models and their application for the modeling of point and nonpoint water pollution. Chapter 3 will deal with stormwater runoff models for both urban and rural applications. Air quality models will be the subject of Chapter 4, which looks at various approaches to the modelling of major air pollutants.

The next two chapters will look at somewhat different approaches to environmental modelling. Chapter 5 will review the computerized approaches to land capability analysis. Here, the focus is much more on the general allocation of land use rather than on the specific impacts of regional or local decisions on the quality of the environment. Environmental quality is considered but in a more indirect manner. In Chapter 6, the emerging area of hazardous waste disposal, the movement of wastes in surface and groundwater, and their ultimate impacts will be discussed. Here, attention is just beginning to focus on the modelling of these impacts and thus fewer computerized approaches exist. Yet, because of the growing importance of this area, it is felt that some attention should be given to this subject.

For each of the subject areas in question, the chapter begins with a general overview. The nature of the impacts that have been studied, the theoretical underpinnings of the models important to understanding their applications, and the different scales at which analyses might be undertaken will be reviewed. Then, several models or approaches will be discussed in detail in light of the evaluation steps discussed above. Each of these discussions will include a comparison of model requirements, applicability, problems, and instructions for obtaining and running the model. Thus, the book should serve as a guide to the use of environmental models for decision making at various geographic scales for a wide range of topical areas, and as introductory instructions on the use of selected, major models in each subject area.

REFERENCES

Andrews, John F. (1978). "Applications of Some Systems Engineering Concepts and Tools to Water Pollution Control Systems," in *Mathematical Models in Water Pollution Control*, ed. by A. James. New York: John Wiley & Sons.

Bertalanffy, L. V. (1968). *General Systems Theory*. New York: George Brazilier.

Fisch, Oscar and Steven I. Gordon (1980). *Evaluation and Testing of NSF-RANN Sponsored Land-Use Modelling Projects with Ohio as a Test Case*. 7 vols. Columbus, OH: Ohio State University Research Foundation.

Haimes, Yacov Y. (1982). "Modeling of Large Scale systems in a Hierarchical Multiobjective Framework," in *Large Scale Systems*, ed. by Yacov Y. Haimes. New York: North Holland Publishing Company. P. 1–17.

Thomann, Robert V. (1982). "Verification of Water Quality Models," *Journal of the Environmental Engineering Division, Proceedings of the American Society of Civil Engineers*, Vol. 108, No. EE5, p. 923–940.

2
Water Quality Models

INTRODUCTION

Applications of water quality models are a required part of many local and regional decisions. The location of major industrial facilities must make use of water quality analyses. Under the Clean Water Act of 1977 (for a review of its provisions see *Environmental Law Reporter,* 8 ELR 10010, January, 1978), and its National Pollutant Discharge Elimination System (NPDES), all public and private facilities that discharge wastes into water bodies must apply for a discharge permit. This application must show the nature and quantity of pollutants which will be discharged at any given site. The application is evaluated in terms of meeting the effluent standards for discharges for that type of plant and the ambient or in-stream standards for water quality. Thus, in the selection of a plant location, the design of plant manufacturing and waste processing facilities (if any), or the expansion of an existing industrial facility, some analysis of the water quality impacts must be made.

A related application of water quality models comes from the location and design of municipal waste treatment facilities. Under the Clean Water Act, several sections require planning for and provide partial Federal funding of waste treatment facilities. The location of these plants and their design vis-à-vis treatment level, plant size, and projected effluents must be subjected to an in-depth analysis of the water quality impacts.

Not only are these so-called point source effluents subject to regulation (i.e. effluents coming out of an identifiable pipe) but nonpoint sources of pollution are also reviewed (i.e. storm water runoff from both urban and rural areas). The projection of the impacts of land use decisions on runoff quantity and quality thus becomes the subject of computer analysis.

THEORETICAL OVERVIEW

Although it is beyond the scope of this book to review at length the theory behind each of the models being discussed, it is important at least to briefly overview some of the theoretical underpinnings of the common models in order to understand the limitations of the computer models.

Very simplistically, water pollutants can be divided into two major groups: conservative and non-conservative pollutants. This division is not a reflection of voting behavior but rather is a measure of the degree to which the pollutant changes chemical states within the water body. Those pollutants that predominantly remain chemically stable, that is, are not transformed to a different chemical state, are said to be *conservative*. In reality, almost all pollutants change form to some degree but their reaction rates are sometimes so slow that they can be ignored for the purposes of the modelling we will be discussing. *Non-conservative* pollutants are converted to new forms at significant reaction rates so that these reactions must be incorporated into any model of the process.

Figure 2-1 illustrates the general process of stream modelling. Each stream segment i is distinguished by having a unique set of physical properties which might affect the dispersion and degradation of pollutants. These include depth, width, slope, velocity, and discharge rate, which are all interrelated; type of stream bed; and turbulence of the flow. Each stream segment also may have a set of natural and human pollutant contributions. These may be from point sources, P_i, or non-point, area sources, A_i. How much of the pollutants from each upstream segment reaches the downstream segment depends upon the rate of decomposition (for non-conservative pollutants) and the dilution of the pollutants by the stream water. This may be thought of as a transfer function, T_i, which depends in turn on stream characteristics, rate of pollution, amount of dilution, and the chemical processes involved. All of this may simply be summarized as follows:

$$C_i = f(P_i, A_i, T_{i-1})$$

$$T_{i-1} = f(S_i, R_i)$$

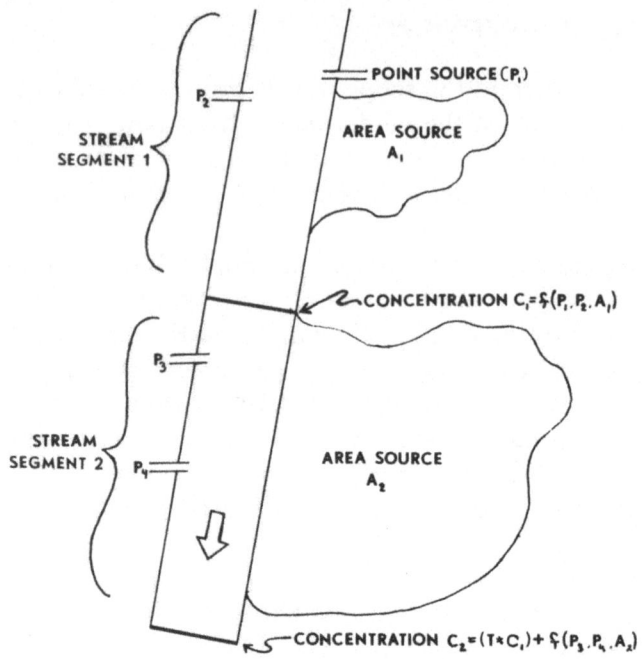

Figure 2-1. Stream modeling.

where

C_i = the concentration of pollutant C in stream segment i;
P_i = the point source contribution of pollutants in segment i;
A_i = the area source contribution of pollutants in segment i;
T_{i-i} = the transfer function of pollutants from the upstream segment;
S_i = the stream parameters affecting the transfer rate;
R_i = the chemical reactions affecting the transfer.

In applying this general model, a number of specific choices can be made with regard to the nature of system representation:

1. A *steady state model* is a model which assumes time-invariant inputs, measurements, and states (Beck, 1978, p. 139).
2. A *dynamic model* allows the explicit changes of the variables and states of the system through time (Sisson, 1975, p. 10).

3. A *deterministic model* assumes that the value of all variables can be computed exactly and that all values are known.
4. A *probabilistic model* assumes that at least some parameters are unpredictable, usually within some range or known distribution and with some probability of occurrence.

One might also distinguish between theoretical models, derived from first principles, and empirical models, derived using statistical techniques. However, in the environmental area there are few models that are entirely of one type or the other. Rather, most models are a combination of theoretical principles and empirically derived relationships. Other classes of models can be specified but these descriptors should be sufficient for our purposes in the discussion of water quality models.

One other distinction among types of water quality models that should be made is between (1) models focusing primarily on the physical or chemical *processes* in a natural body of water which result in a particular pollution level, and (2) models focusing on the *costs* of pollution abatement strategies or plant designs. In actual practice, both types of models will be important in making decisions. However, the cost models that are available tend to be oriented toward unique, local situations where physical design is constrained and(or) local construction and operation costs are used. This makes transfer of these models to other uses more difficult. For these reasons, the models on which we will focus are all of the physical-chemical type.

STREETER-PHELPS MODELS: DISSOLVED OXYGEN IN A RIVER

The most developed subset of water quality models are the models of dissolved oxygen in streams and estuaries. There are several reasons for this. Dissolved oxygen is one of the most critical ecological indicators of the health of a stream. Almost all stream biota require oxygen for their survival. The decomposition of organic wastes within the stream also requires oxygen. If the supply of oxygen becomes depleted, the consequences then include the death of a large portion of the stream biota as well as the potential for the creation of septic conditions which produce unattractive, odoriferous, and unusable streams.

Historically, dissolved oxygen has also been studied the most because the visible pollution from organic wastes in urban, industrialized areas are much more obvious than the chemical pollutants which frequently coexist. In most cases, the volume of organic waste discharges far exceeds that for other pollutants, making these wastes the subject of the earliest attempts to improve water quality. Of course, many of the organic chemical pollutants that we worry about today have not been in existence for a very long time. Then too, we understand much less about the chemical fate of these complex compounds, partly because they are so new and partly because they are so expensive and complicated to detect. Given adversities of this magnitude, the general reaction of much of the scientific world is to take the easy way out—study the better known, more accessible pollutants of greater magnitudes. It is for all these serious, practical, and pedantic reasons that dissolved oxygen and the organic wastes leading to its depletion have been the focus of much water pollution modelling work.

The seminal work on dissolved oxygen (DO) was performed by Streeter and Phelps (1925) and has led to a class of DO models bearing their names. The basic principles underlying their model involve the rate of depletion of oxygen from the water—the *deoxygenation rate,* and the rate of replacement of oxygen from the atmosphere—the *reaeration rate.* In particular, the original Streeter and Phelps model postulated that the rate of deoxygenation is directly related to the quantity of organic wastes to be decomposed in the water, while the reaeration rate is a function of stream characteristics affecting oxygen exchange with the atmosphere. These processes are also impacted by the temperature of the water since the ability of the water to hold oxygen, called the *saturation level,* is inversely related to its temperature.

Figure 2-2 illustrates the simplest case of a stream with a single point source discharge causing a depletion or sag in the oxygen level. The total amount of dissolved oxygen present in the water before a discharge occurs is indicated here as the upper asymptote, the saturation level. This level is governed by the water temperature. The

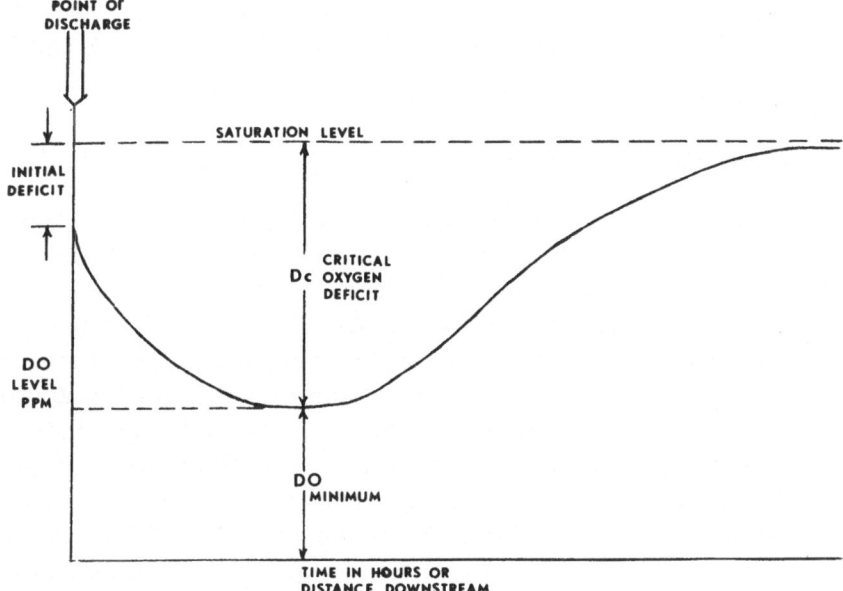

Figure 2-2. Dissolved oxygen sag curve.

actual DO level never quite reaches this level because there is always some natural organic waste in the water and the stream biota use some oxygen for respiration. The physical limits of DO are 0.0 at the lower end and 14.6 at the upper end.

(The exception to this upper limit comes as a result of a condition called *supersaturation*. This may occur immediately downstream of an unusual amount of turbulence—for example, below a dam—that captures an extraordinary amount of DO from the atmosphere. Supersaturation may also occur in streams with algae problems, especially during the day when the algae release a large amount of oxygen from photosynthesis. This condition is offset by large respiration losses at night.)

Going back to Figure 2-2, we see that at the point of discharge there is an immediate deficit in DO. This is caused by the mixing of stream water with a volume of effluent water having a very low DO level. The initial DO deficit can be calculated as follows:

$$DO_m = \frac{(DO_s \times Q_s + DO_e \times Q_e)}{(Q_s + Q_e)}$$

where DO_m, DO_s and DO_e are the DO levels of the mixture, the stream, and the effluent at the time mixing takes place; and Q_s and Q_e are the water volumes from the stream and effluent respectively.

Oxygen vs. Wastes

As time goes on, or one moves downstream from the effluent source, the level of oxygen in the stream begins to decline. Here, the deoxygenation rate, related to the amount of organic wastes in the water, exceeds the reaeration rate. Reaeration is related, in turn, to the physical attributes of the stream, to the oxygen deficit (which affects the rate of diffusion of oxygen from the atmosphere), and to any artificial structures (bridge abutments, dams) that might increase the turbulence of the stream and therefore the exchange of oxygen with the atmosphere. Through each increment of distance, both processes proceed until DO reaches a minimum level. This is the most critical DO level because it represents the level which biota must tolerate in order to survive in this stream reach. The curve then begins to rise as organic wastes are decomposed to such a degree that reaeration exceeds deoxygenation. After all of the wastes have been decomposed, the DO level again approaches the saturation level.

Actually, there are two segments to the oxygen demands of wastes—a *carbonaceous demand* and a *nitrogenous demand*. The carbonaceous demand is generally met first followed by a secondary, nitrogenous demand. The original Streeter-Phelps model does not distinguish between these processes but more recent efforts take them into account (See Fair et al., 1971; Wang et al., 1979; Hughto and Schreiber, 1982; Beck, 1978).

The Streeter and Phelps model as developed by them and others represent this process by the following differential equation (Streeter and Phelps, 1925; Clark et al., 1971):

$$\frac{dD}{dt} = K_1 L - K_2 D$$

where

$D =$ oxygen deficit (mg/l) at the downstream distance X (m);
$t =$ time (days);

K_1 = the deoxygenation coefficient;
L = oxidizable organic matter (mg/l ultimate BOD) present at any time (t);
K_1L = rate of deoxygenation (mg/l-day);
K_2D = rate of reaeration or reoxygenation (mg/l-day).

BOD is a measure of the organic waste load of the stream which is determined via a laboratory test. The Streeter-Phelps method is thought to be reliable as long as K_2 and K_1 are determinable. However, in many circumstances the values are unknown and are then taken from standard tables relating to stream conditions. If the assumptions implied by these tables do not hold, the model becomes unreliable.

Further Impacts on Streeter-Phelps Model

Other types of conditions also affect the reliability of this model. The model does not have a very reliable method of incorporating the consideration of non-point sources of water pollution. Some researchers have gotten around this by introducing an "equivalent" point source of pollution at various locations along the stream. The model in the form given above does not account for oxygen lost due to the respiration of biota or that produced by the photosynthesis of plants. Various adjustment terms have been introduced to correct for these factors. Finally, the model does not reflect the interaction between the stream benthic (or bottom) deposits and organic waste loads. During times of low flow, organic wastes may be deposited on the bottom and thus taken out of the water relative to the decomposition process. In times of increased flow, these wastes may be stirred up and increase the actual organic waste load on the stream at that time. No reliable estimate of the impacts of these loads on the accuracy of DO models appears to be available although a number of standard "correction" factors can be employed.

Several stream characteristics affect the value of the deoxygenation and reaeration rates. The deoxygenation rate is related to the water temperature since the rate of chemical reaction increases with the temperature. This rate is also related to the reaeration rate in that the availability of oxygen to allow decomposition to proceed is limited by the amount of oxygen present. The reaeration rates are affected mostly by the amount of mixing between the water surface and the atmosphere. As such, any physical condition which increases the turbulence of the water will increase the reaeration rate. Rougher stream beds, man-made structures such as bridges and dams, and stream slope will all contribute to higher reaeration rates.

One other attribute of these models should be pointed out. Although the model appears to account for time, it is not dynamic. Rather, it is a steady state model which assumes no time changes in effluent levels, deoxygenation and reaeration rates, stream discharge levels, or stream conditions. As such, it is subject to many possible deviations from actual DO values. Nevertheless, it is still the model that is used most frequently in water quality analysis for general planning purposes.

Computerizing Streeter-Phelps

Several computerized versions of the Streeter-Phelps formulation are available, both for microcomputers and mainframe systems. Table 2-1 indicates the input data requirements and outputs of these models. As one can see, the model requires a good deal of information about the stream and about each effluent source. Data on the effluent sources are generally available through the USEPA or their state counterparts. However, the data may or may not be in computer compatible format, making data entry and acquisition very costly. Even if data are available in computerized form, the reliability may be questionable. Some firms may report only their average effluent outflows while others provide more details of average and peak flows. Even if peak flows are reported, their frequency of occurrence may be unknown. Thus, deriving a representation of the range of effluent flows over time for a particular stream will frequently be a tedious task and require some data manipulation and some guess work.

Data for the streams themselves may be equally problematic. Water companies, sewage treatment plants, private corporations, and various public agencies (EPA, health departments, U.S. Geological Survey, River Basin Commissions, quasi-public basin organizations) all collect data. However, each differs in its sampling network,

Table 2-1. Data Requirements and Outputs from Streeter-Phelps Dissolved Oxygen Models

Inputs	Outputs
For each stream segment:	For each stream segment:
Discharge rate	Oxygen deficit at any point
Stream bed characteristics	DO level at any point
Temperature	Minimum DO level
Saturation level (related to temperature)	Point of minimum DO
Reaeration coefficient (related to stream)	
BOD of stream	
Deoxygenation coefficient	
Stream velocity	
Average stream depth	
Length of segment	
Slope of segment	
For each pollution source (by segment):	
Effluent flow	
Effluent temperature	
BOD of effluent (sometimes divided by type)	
DO in effluent	
Temperature of effluent	

frequency of collection, length of record, and types of variables sampled. Some use "grab" samples, those collected at one time by a technician in the field, while others use continuous monitors. Large segments of data may be missing for reasons such as a broken monitor, ill technician, treatment plant not working as required, and so on. Here again, assembly of the required data can become a very extensive task. For smaller stream tributaries, there may be no monitoring data at all.

Fortunately, many larger stream systems have in place various data bases and modelling efforts. Thus, the data collection and analysis tasks can be short-circuited in these cases. Given that the data can be found, the Streeter-Phelps model requires the analyst to simulate the conditions existing for average and peak effluent loads, for representative times of the year (for example, the summer would produce the lowest DO levels because of temperature conditions), and to estimate the frequency of occurrence of various conditions. In this way, one can establish a base of conditions showing where DO is a problem, falling below standards, how often this occurs; and who are the major polluters in the basin. It is only through the use of such a base that various policy decisions can be made. The major policy options that might be examined using such a model are

1. Locations for new point source polluters;
2. Treatment levels for new and existing point source polluters;
3. The need for new public and private waste treatment facilities;
4. The impacts of alternative decisions for the above.

Aside from technical limitations, this type of model can become very cumbersome to use because of its extensive data requirements and the large number of runs that are needed to examine the results of each potential policy. Because the model calculates DO levels on a scale of days, many simulations must be used to formulate a representation of annual events. This requirement is coupled with a large number of policy scenarios and the technical problems with trying to incorporate non-point sources, benthic demand, and so on, the model will likely become very difficult to use. Nevertheless, it forms the basis of the majority of simulations for the initial planning for water quality.

Obtaining Computer Models for Streeter-Phelps

The equations required to perform the analysis of the Streeter-Phelps model are relatively easy to program directly. There are also several computer models, in Basic and Fortran, that have been published in the literature and thus could be used on either a microcomputer or a mainframe system. Wang et al. (1979) list a fortran program of a DO model. Hughto and Schreiber (1982) have a Basic program which performs similar calculations. For those wishing to write their own simple program, some good examples of the Streeter-Phelps formulation are given in Clark et al. (1971, p. 274–279).

For some larger river systems, extensive modelling has already been completed. Access to these models is sometimes possible via contacting the relevant administrative agency (EPA, River Basin Commission, Regional Planning Agency). However, these models are not always transportable to other systems; they may be difficult to use because of their size and complexity; and frequently they are not documented well enough to allow use by outsiders. Another approach to analyzing such complex situations is introduced in the following section.

A STATISTICAL APPROACH TO DISSOLVED OXYGEN MODELLING

A number of authors have suggested the use of statistical techniques as an alternative approach to environmental modelling. These approaches use historical data on environmental pollution levels, physical and chemical parameters, and loading rates to develop statistical relationships that represent the reaction of the system to parameter changes. Then, forecasts are made of future conditions and used to project the impacts of alternative policies. Several implicit and explicit assumptions must be made in using such an approach to water quality modelling:

1. There must be a relatively consistent and lengthy historical record of water quality, effluent, land use, and stream data.
2. There must be sufficient resources to put the data in computer-readable form or the data must already be in this form.
3. The past record is representative of the range of events that have occurred and will occur in the future.
4. There are some known or hypothesized physical relationships among the variables.

Several authors have utilized statistical techniques in the modelling of water quality. Nitrogen, phosphorous, and suspended solids in New York State rivers were related to land use patterns using a multiple regression technique (Haith, 1976). A statistical model of streamflow, BOD, and DO was validated by Stochastics Inc. for the Potomac Estuary (U.S. Army Corps of Engineers, 1974). Carey et al. (1972) formulated a model with similar outputs for the Raritan and Passaic Rivers in New Jersey while Gordon and Fromuth (1981) followed the New Jersey study's approach for a model of the Great Miami River in Ohio. Beck (1978) provides a comprehensive overview of various approaches.

Setting Up a Statistical Model

Before we focus on an explicit example, it might be useful to review the general process involved in the statistical modelling of water quality (particularly DO). First one must determine, for the river basin in question, that the relevant available data covers a sufficiently long period of time. This means not only having a period of record of at least three to five years but also having a relatively complete dataset for the entire period. The dataset must then be "cleaned." Depending on the dataset's format, this may involve the following steps:

1. Code and key the water quality data into the computer. This process will involve creating or adopting a numbering system for gauging and monitoring stations, determining which data are of relevance, copying the data from their current, written form to code sheets, and having them keypunched to cards, tape, or disk. If the data are already in computerized form, some effort must be expended in obtaining a copy of the data, the documentation for the computer files, and reading the information to confirm that one understands the documentation correctly. Frequently, the data are on tape. On larger systems, this does not create problems; but on smaller ones, options to read and translate certain types of tapes may not exist. See Appendix A for a brief review of these potential problems.
2. Run listings and simple frequency distributions on the data to determine the number and frequency of missing values, the locations of stations where data are missing, the type of data missing, and the range, mean, and other descriptive statistics. This process should help determine whether or not the data are sufficient to build a computer model. It will also show, in descriptive terms, where current and potential water quality problems in the basin occur. It is important to know this to determine whether future model outputs make sense.
3. Gather information on stream parameters in the basin. These will be the physical parameters that are important to DO levels, such as slope, turbulence, stream bed, artificial stream impediments, and so forth, and the basin boundaries. Most of these data must be gathered in the field or measured on standard U.S. Geological

Survey maps. Again, these data may have been tabulated for some basins.

4. Encode data on the effluents emitted into the streams. Coding may have to be done especially if several input sources of data are used. If information on non-point sources are to be considered, land use data may be required. If no recent historical data exist, which is likely, remote sensing and field survey techniques may have to be used at great expense.

5. Formulate the structure of the model in question. What are the hypothesized statistical relationships? Here, other, theoretical efforts are important as a guide.

6. Input the data to a standard statistical package to derive the empirical models. The most flexible package for this purpose is the Statistical Analysis System (SAS), 1976.

7. Validate the model and use it for policy analysis.

Some Applications of a Statistical DO Model

Keeping these steps in mind, let us now examine a statistical DO model which has been employed in several situations. The model in question is based on the work of Thomann (1967) and has been applied to three river basins by two different research teams (Carey et al., 1972, and Gordon and Fromuth, 1981). In the latter case, data on the Great Miami River were used to create a DO model. Fortunately, in this case a long term, consistent dataset on the DO levels in the river was available from the Miami Conservancy District on computer tape. Data on 24 sampling stations were compiled and combined to create a dataset showing the following:

1. Biweekly dissolved oxygen data for monitoring stations along the main stem of the Great Miami River.

2. Point BOD loads for all major sources including industries and sewage treatment plants. These were obtained from the conservancy district and the Ohio Environmental Protection Agency.

3. Slope of the stream bed between monitoring stations.

4. The major geologic provinces within the basin.

5. Urban land use data for land along the main stems as a representation of potential non-point pollution.

6. Rural land use and soils data as a proxy for rural non-point source pollution.
7. Streamflow and cross-section data for each DO monitoring station.

Biweekly DO data were used because this provided the most consistent time frame for which data were available. Although some daily readings were available for some stations, others only had weekly or biweekly grab samples.

A number of model equations were then tested using multiple linear regression. Here, the mean annual DO level was predicted as a function of such variables as point source BOD, land cover and land use, geologic province, etc. The equation can be represented as:

$$DO_{\mu j} = \sum_i \alpha_{ij} P_{ij} + \sum_k b_{kj} F_{kj} + \epsilon$$

where

$DO_{\mu j}$ = mean dissolved oxygen level for station j (mg/l);
P_{ij} = a set of i pollution variables related to mean DO at station j;
F_{jk} = a set of k stream variables related to mean DO at station j;
ϵ = an error term;

A number of formulations were tested with the final one being selected on the basis of statistical significance, theoretical meaning, and accuracy.

The second model equation assumes that DO variation over the period of a year is a trigonometric function of time. Figure 2-3 represents this graphically. The annual variation around DO_μ is a curve of amplitude A_j that is a function of time. Carey et al. related this amplitude to the dissolved oxygen variance for a given station:

$$A_j = (2\sigma_j^2)^{1/2}$$

where

A_j = amplitude of the dissolved oxygen curve for station j (mg/l);
σ_j^2 = variance of bi-weekly dissolved oxygen readings for station j around $DO_{\mu j}$ (mg/l).

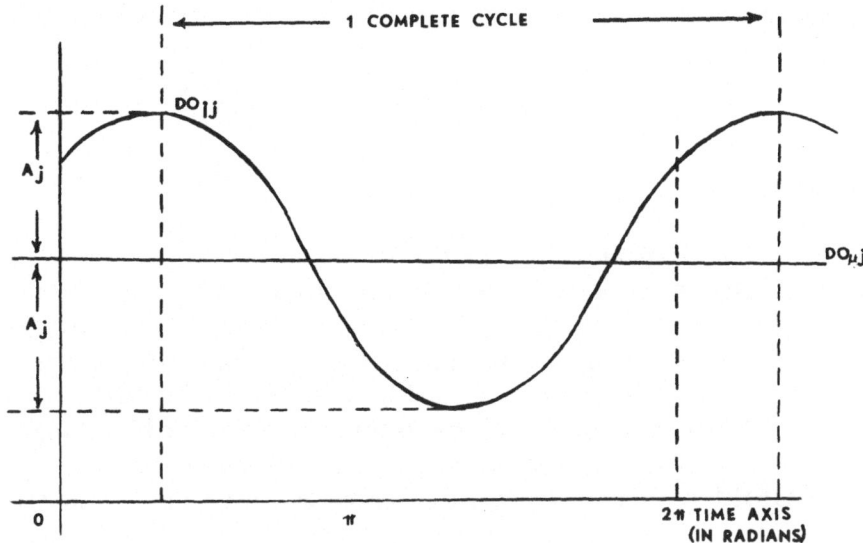

Figure 2-3. Graphical representation of annual DO cycle. (Source: Gordon and Formuth, 1981. Reprinted with permission.)

The biweekly dissolved oxygen level is simulated by the third model equation represented in Figure 2-3:

$$DO_{ij} = DO_{\mu j} + A_j \cos\left(\frac{\pi(T_{ij})}{13} - \theta_j\right)$$

where

DO_{ij} = the ith biweekly dissolved oxygen level for station j;

$DO_{\mu j}$ = mean DO level from equation (1);

A_j = amplitude of the curve from equation (3);

T_{ij} = the biweekly period (fortnight) associated with the ith dissolved oxygen level at station j. Values range from 1 to 26;

θ_j = the phase constant for station j which permits DO_{ij} to reach a maximum value at the correct time. The value of θ is estimated empirically.

The model was tested using data for the Great Miami River for 1973–1975 and validated with data for 1976. Four variables were found to be the major causes of variation in mean annual DO levels:

$$DO_{\mu j} = 10.317 + 0.173 \, BODU_j - 0.0617 \, BODS_j \\ - 0.399 \, FOREST_j + 0.684 \, COMM_j$$

where

$DO_{\mu j}$ = mean annual DO level for station j;

$BODU_j$ = mean daily point source BOD load transferred to station j from the upstream station $j - 1$ (10^3 lb/day)

$BODS_j$ = the mean daily point BOD generated in the stream segment represented by station j (10^3 lb/day);

$FOREST_j$ = land in reach j that is forested plus forest land that is transferred from upstream (10^3 acres);

$COMM_j$ = land in reach j in commercial use plus commercial land from upstream (10^3 acres).

The equation for the amplitude is as follows:

$$A_j = 6.234 - 0.348 \, DO_{\mu j}$$

The transfer function for both point and non-point BOD (using the land use proxy variables) is calculated using the step function related to stream slope derived by Carey et al. and shown in Table 2-2. This function is based on the fact that the efficiency of water transfer is lower in higher slope streams. The reason for this is that such streams tend to have rockier bottoms and more turbulent flow. Thus the downstream progress of water is hampered, allowing more time for the degradation of organic wastes. In lower slope streams,

Table 2-2. Step Function of BOD Passed Downstream under Different Slope Conditions

Status of Stream Flow	Slope	BOD Accumulated at Downstream Station	Remarks
Fast	$S < 2.5$	100%	BOD transported: no assimilation
Medium	$2.5 < S < 5$	50%	Some BOD assimilated
Slow	$S > 5$	0%	All BOD assimilated

Source: Carey et al. (1972), p. 156.

the bottom tends to be of finer materials—sand and silt—producing more laminar flow and thus a more efficient downstream movement. This allows less time for the degradation of the organic wastes per unit of distance downstream. For these reasons, high slope streams (greater than 5%) tend to assimilate all the BOD. Intermediate slopes assimilate some and transfer some of the BOD. Between 2.5 and 5% the model assumes a 50% transfer rate while streams with slopes less than 2.5% are assumed to transfer all of the BOD. Sensitivity testing has shown that the model is not sensitive to reasonable changes in these parameters. However, as we will see, this assumption can produce a trap for the analyst in certain kinds of policy analysis.

The model has been validated using regression analysis and shown to be an excellent predictor of DO. This is shown in Table 2-3. The

Table 2-3. Standard Error of Estimate, Mean Error of Estimate, and R^2 without Intercept for 1973-75 for Predicted Biweekly DO Levels

Station I.D.	River[2] Mile	Type of Station	No. of OBS.[3]	Standard Error (MG/L)	Mean Error (MG/L)	R^2 Without Intercept	Mean Observed DO (MG/L)
1	111.90	G[1]	31	1.5	1.2	.99	9.1
2	108.04	G	33	2.2	1.5	.95	9.6
4	99.00	M[2]	66	1.4	1.0	.98	10.2
5	93.98	G	31	2.5	1.6	.93	9.3
6	92.45	G	26	2.2	1.7	.97	11.0
7	90.87	G	34	1.5	1.2	.97	9.0
8	89.45	G	31	2.2	1.7	.95	9.9
10	82.68	G	31	2.5	1.9	.94	10.0
11	77.96	M	73	1.3	1.0	.99	10.8
14	76.36	G	44	1.7	1.3	.97	9.7
16	72.91	G	33	1.3	1.1	.98	9.4
17	72.72	G	31	1.3	1.2	.99	10.2
18	69.00	G	48	1.9	1.6	.96	9.0
20	66.43	G	25	2.0	1.5	.94	8.3
21	65.75	M	69	1.6	1.2	.97	8.6
22	64.34	M	63	2.0	1.2	.94	8.1
24	63.82	M	58	1.4	1.0	.97	8.0

[1]Grab Station: sampling frequency less than once per week.

[2]Monitor Station: DO level recorded at 2 hour intervals.

[3]An observation is the average DO level for a biweekly period. In the case of monitor data, an observation represents the mean of two weeks of 2-hourly DO readings. In the case of grab sampling stations, a single observation may be the only DO measurement for the biweekly period.

Source: Gordon and Fromuth, 1981.

R^2 values are between .94 and .99 for all stations with 68% of the stations having mean standard errors of less than 2.0 ppm. This indicates that for general planning purposes the model performs quite well as a predictor of impacts of both point and non-point organic wastes on the DO level.

The computer version of this model is relatively straightforward. One must use a statistical program to generate the equations from the input data for the stream basin in question. Given the final regression equations, one can use the same statistical program to input the variables and derive the simulated DO values. The SPSS, BMDP, and SAS statistical packages, available as standard statistical routines on many mainframe systems, have the capabilities necessary to run this model. Similar packages for microcomputers are also becoming available.

COMPARISON OF THE DO MODELS

It may be useful to compare the two DO models discussed so far in terms of their assumptions, policy analysis capabilities, and potential errors. Neither of the models are dynamic with respect to time. Both assume that all of the parameters are invariant with respect to time except for the spatial variation of the implied time parameter.

Spatial variation is also limited to those sample points represented by the stream segment break points or monitoring stations for which data are available. No other spatial breakdown of the data is possible.

Although both models can simulate point source load impacts, the Streeter-Phelps model handles non-point source loads with great difficulty. Non-point source loads are at least accounted for by proxy variables in the statistical model. The time scale of the two models also differs. The Streeter-Phelps implies a time scale of days while that for the statistical model, at least as used here, is biweekly. Both models demand a good deal of data although the statistical model requires more in the sense of a consistent and complete dataset over a significantly longer time period.

Given these characteristics, the advantages and disadvantages for each of the models follow. The Streeter-Phelps model is widely accepted and used and requires much less data in order to be made operational. For simple stream situations (i.e. not too many pollu-ters, not a large non-point source problem, interest in the impacts of one or two new point sources at one or two locations) Streeter-Phelps and models related to it are probably preferred. However, when the situation becomes more complicated, this model becomes very cumbersome. The reason for this is the level of detail relative to the implied time and spatial scale. The Streeter-Phelps model implies a large number of physically defined stream segments and simulates the impacts on these segments in terms of hours or days.

In order to arrive at a decision regarding the overall policy implications of large scale decisions, one must examine a tremendous amount of computer output. For example, if one were trying to decide what the treatment levels were to be for all major industrial and municipal waste treatment facilities along a river system, one would need to simulate the impacts through a relatively arduous process requiring the processing and analysis of data for each segment, for many time periods, and for many potential treatment levels. With the statistical DO model, given that the data are available, both point and non-point pollutants can be simulated. Because of the higher level of generality, the analysis of long term trends and major policy decisions becomes much simpler. Tables 2-4 and 2-5 illustrate these points by showing the typical output from both types of models.

The major drawbacks to both models come from the nature of the assumptions that are made with each. Both models assume a steady state system. This precludes consideration of real variations in the quantity of oxygen-demanding wastes, flow changes in the river system caused not only by seasonal variations in inflow but also by withdrawals for municipal and industrial use, changes in streambed conditions, and variations in temperature. Neither model handles the effects of other parameters very well. Extremes caused by algal blooms, bacteriacidal chemicals that deplete the natural assimilative capacity of the stream, and artificial reaeration by manmade struc-tures will cause real and sometimes dramatic changes in dissolved oxygen which will be completely missed by both models.

The Streeter-Phelps model has some additional problems of note. Accurate determination of the deoxygenation and reaeration con-stants is frequently very difficult. Compensatory corrections for benthic demand, non-point sources, and respiration have been

Table 2-4. Representative Output from Streeter-Phelps DO Model

Distance (miles)	DO Deficit (mg/l)	DO Level (mg/l)
Stream Segment 1		
0.00	2.40	7.6
0.20	2.38	7.6
0.40	2.38	7.5
0.60	2.36	7.5
0.80	2.36	7.5
1.00	2.36	7.5
1.20	2.34	7.5
Stream Segment 2		
1.40	2.34	7.5
1.60	5.90	3.9
1.80	6.00	3.8
2.00	6.20	3.8
2.20	6.30	3.7
2.40	6.35	3.6
2.60	6.35	3.6

developed but are frequently applied without any validation or measurement of model accuracy.

The statistical dissolved oxygen model also has some idiosyncracies. Validation is somewhat easier in that it follows from the statistical analysis which must be performed. Thus, the model user has a better idea of overall accuracy. However, certain types of conditions produce erroneous results. Given the assumptions about assimilation capacity as it relates to stream slope, one can create a condition where, regardless of the amount of new point source BOD which is added at a station, the BOD will completely disappear by the time it arrives at the next station. This of course is impossible given the physical limits to stream assimilation capacity. But the model does not reflect this upper limit since it is linearly based with no upper asymptote. The model, however, could be altered to reflect such limitations.

Even given these reservations, both models are useful for general planning and impact analysis work. The potential site and regional impacts of decisions regarding new plant locations, treatment levels, and land use decisions can be simulated using these models. In this

Table 2.5. Representative Output from Statistical DO' Model

DO (I,J), STATION 172 PLAN AP*

YEAR 1975

0	11.4	7.4	7	9.2	5.2	14	3.8	−0.2	21	7.3	3.3
1	11.7	7.7	8	8.3	4.3	15	3.7	−0.3	22	8.3	4.3
2	11.8	7.8	9	7.3	3.3	16	3.8	−0.2	23	9.2	5.2
3	11.7	7.7	10	6.3	2.3	17	4.1	0.1	24	10.1	6.1
4	11.4	7.4	11	5.4	1.4	18	4.7	0.7	25	10.8	6.8
5	10.8	6.8	12	4.7	0.7	19	5.4	1.4			
6	10.1	6.1	13	4.1	0.1	20	6.3	2.3			

YEAR 1980

0	11.3	7.3	7	9.1	5.1	14	3.7	−0.3	21	7.2	3.2
1	11.7	7.7	8	8.2	4.2	15	3.6	−0.4	22	8.2	4.2
2	11.8	7.8	9	7.2	3.2	16	3.7	−0.3	23	9.1	5.1
3	11.7	7.7	10	6.2	2.2	17	4.0	0.0	24	10.0	6.0
4	11.3	7.3	11	5.3	1.3	18	4.6	0.6	25	10.8	6.8
5	10.8	6.8	12	4.6	0.6	19	5.3	1.3			
6	10.0	6.0	13	4.0	0.0	20	6.2	2.2			

YEAR 1985

0	11.3	7.3	7	9.1	5.1	14	3.6	−0.4	21	7.1	3.1
1	11.6	7.6	8	8.1	4.1	15	3.4	−0.6	22	8.1	4.1
2	11.8	7.8	9	7.1	3.1	16	3.6	−0.4	23	9.1	5.1
3	11.6	7.6	10	6.1	2.1	17	3.9	−0.1	24	10.0	6.0
4	11.3	7.3	11	5.2	1.2	18	4.5	0.5	25	10.7	6.7
5	10.7	6.7	12	4.5	0.5	19	5.2	1.2			
6	10.0	6.0	13	3.9	−0.1	20	6.1	2.1			

DO (I,J), STATION 172 PLAN AB

YEAR 1975

0	11.5	7.5	7	9.4	5.4	14	4.1	0.1	21	7.5	3.5
1	11.8	7.8	8	8.4	4.4	15	4.0	−0.0	22	8.4	4.4
2	11.9	7.9	9	7.5	3.5	16	4.1	0.1	23	9.4	5.4
3	11.8	7.8	10	6.5	2.5	17	4.4	0.4	24	10.2	6.2
4	11.5	7.5	11	5.7	1.7	18	5.0	1.0	25	10.9	6.9
5	10.9	6.9	12	5.0	1.0	19	5.7	1.7			
6	10.2	6.2	13	4.4	0.4	20	6.5	2.5			

*Values are reading number, DO level, and DO minus 4.0 ppm standard for various locations, years, and projected treatment options.

Table 2.5. Representative Output from Statistical DO Model (cont.)

DO (I,J), STATION 172 PLAN AB (cont.)

YEAR 1980

0	11.4	7.4	7	9.3	5.3	14	4.0	−0.0	21	7.4	3.4
1	11.8	7.8	8	8.4	4.4	15	3.9	−0.1	22	8.4	4.4
2	11.9	7.9	9	7.4	3.4	16	4.0	−0.0	23	9.3	5.3
3	11.8	7.8	10	6.4	2.4	17	4.3	0.3	24	10.2	6.2
4	11.4	7.4	11	5.6	1.6	18	4.9	0.9	25	10.9	6.9
5	10.9	6.9	12	4.9	0.9	19	5.6	1.6			
6	10.2	6.2	13	4.3	0.3	20	6.5	2.5			

YEAR 1985

0	11.4	7.4	7	9.2	5.2	14	3.9	−0.1	21	7.3	3.3
1	11.7	7.7	8	8.3	4.3	15	3.7	−0.3	22	8.3	4.3
2	11.9	7.9	9	7.3	3.3	16	3.9	−0.1	23	9.2	5.2
3	11.7	7.7	10	6.4	2.4	17	4.2	0.2	24	10.1	6.1
4	11.4	7.4	11	5.5	1.5	18	4.8	0.8	25	10.8	6.8
5	10.8	6.8	12	4.8	0.8	19	5.5	1.5			
6	10.1	6.1	13	4.2	0.2	20	6.4	2.4			

way, a general strategy can be developed and then tested in more detail, if necessary, using more accurate models.

WATER QUALITY MODELS EMBEDDED IN OTHER MODELLING EFFORTS

A number of other water quality models for other substances, both conservative and non-conservative, are available in the form of computer programs. A listing of these is provided in the bibliography at the end of the book.

Besides these models, several models concerned primarily with the assessment of other aspects of water resources management have embedded water quality components. A few are stormwater models, discussed further in the next chapter. It appears to be more appropriate to discuss the water quality aspects here. The purpose of these models is to provide a relative measure of the pollution impacts of stormwater rather than to trace the instream impacts of the pollutants. Thus, the models rely on various theoretical and empirical functions which calculate a "pollutograph," or time distribution of pollutants delivered with the stormwater entering the receiving

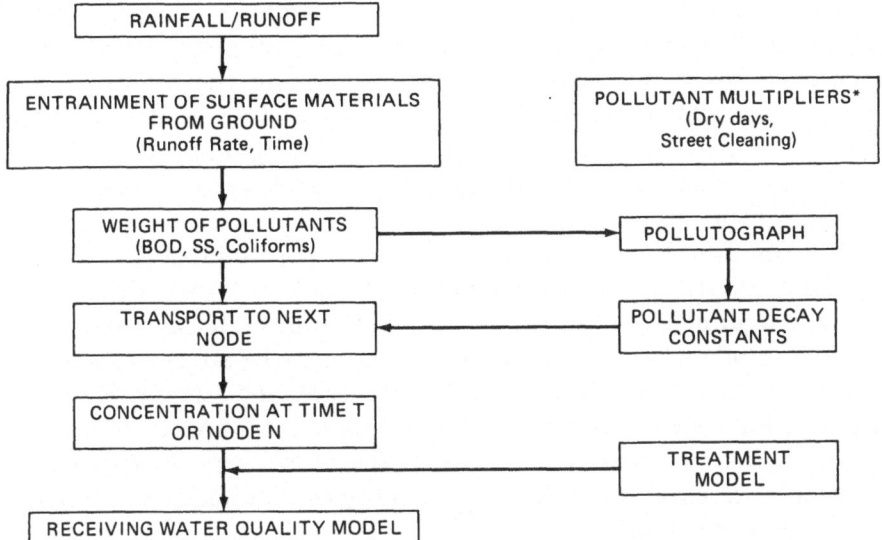

Figure 2-4. Pollution calculation flows in SWMM.

stream. Other models, such as Streeter-Phelps, would then need to be used to calculate the instream impacts of these pollutants.

One example of such an effort is embedded in the Storm Water Management Model (SWMM) formulated under the sponsorship of the USEPA (July 1971). The flow of information in this complex model is represented by Figure 2-4. The pollutant portion of the model begins with input from the subroutines associated with the modeling of the stormwater rainfall-runoff relationship in an urban area. The next subroutine in this model, SFQUAL, uses the hydrograph of surface runoff to produce three pollutographs—for BOD, suspended solids, and coliform bacteria. These pollutants are passed through the sewer system and accumulated until they reach the outflow point where final totals are reported.

The method of performing these calculations is generally illustrated in Figure 2-4. Here one can see that pollutant multipliers are used to predict the amount of pollutants contributed by each urban subarea as represented by a node in the stormwater system. These multipliers are in turn impacted by (1) the number of dry weather days (days between rainstorms) during which pollutants can accumulate and (2) the offsetting of this accumulation through street cleaning. The calculations involved are rather simple in formulation, generally involving linear relationships among the variables and some empirical correction factors allowing a closer match with the limited data used to test the submodels. The weight of each pollutant over time is reported in the form of a pollutograph which is the output of this submodel.

Information is then passed to the transport submodel of SWMM. While being transported to the next node, pollutants will decay, if at all, again along a simple linear decay function. Material is passed to the next node where additional pollutants may be added, and so on, until the stormwater finally reaches the receiving stream. Here, an intermediate step can be used to simulate the impacts of adding various treatment processes. Finally, the model reports a time and spatial representation of the final quantity of pollutants reaching the receiving water body.

The formulation in SWMM is useful in the sense that it provides a comprehensive manner in which to trace the potential impacts of stormwater on water quality. There are several disadvantages to the use of the model however. As will be discussed in more detail in the next chapter, the model is extremely data-demanding and therefore expensive to operationalize for a new area. The water quality estimates in particular are very general, depending on very limited studies in few areas for information on the quantity of pollutants resulting from stormwater runoff. For example, the pollutant multipliers for street runoff are taken primarily from a study done in Chicago by the American Public Works Association (1969). There is certainly a high probability that Chicago is not exactly "your kind of town" when it comes to the pollutographs associated with stormwater runoff. Yet, the necessity of additional local sampling would make use of the model even more expensive and out-of-reach for most projects. Thus, most current users of the model simply assume that places like Chicago are in the right "ball park" for estimates of non-point pollution loadings.

Given these reservations, one must conclude that water quality estimates based on embedded calculations such as those in the SWMM may provide useful, initial estimates of the pollution loadings but that users must be aware of the high level of generalization and the limited empirical validation of such calculations. Some local validation and adjustment of such estimates should therefore be incorporated into any modeling and analysis efforts.

EXAMPLES: MAKING A POLICY DECISION

It may be helpful to work more carefully through the example output for one of the DO models to illustrate how it would be used to make a policy decision. The output of one version of the statistical DO model is given as Table 2-5. The output is formatted to give the following information:

1. The biweekly DO values for each of 26 readings labelled 0 through 25 with the DO values following.
2. The DO values minus the DO standard of 4.0 ppm in the column adjacent to the DO readings. This simple computation allows one to easily spot the locations where DO goes below the standard since these numbers show up with negative values.
3. Output for three simulation years, 1975, 1980, 1985. In this case, the base year of the calculations was 1970 so that these represent three projection years.
4. Output for several alternatives, here labelled as Plan AP and Plan AB representing different assumptions about waste treatment levels.

In order to get these final outputs, the user had to prepare a number of input data for each of the stations in the river basin:

1. Physical data representing stream characteristics including slope, subbasin area, bed characteristics, and so forth.
2. The BOD loads for each station for each simulation year. This entails calculation of the point source BOD from sewage treatment facilities and industries. Such calculations require the user to estimate present loads and then forecast future loads based on assumptions of population growth and waste treatment standards. These calculations must be done exogenous to the model. If the model includes non-point BOD proxies based on land use, these must also be projected and entered in the input data stream.

Given the physical and effluent data, the model calculates the DO averages for each station beginning at the uppermost segment of the basin and working downstream. BOD is passed to the next segment based on the parameters shown in Table 2-2. Then calculation proceeds sequentially to find the amplitude and finally the biweekly DO levels reported on the output.

For the station represented in Table 2-5, the AP scenario projects a BOD load of 1656 pounds in 1975, while that for the AB scenario is 811 pounds in 1975. The BOD climbs in both cases as time proceeds. In actuality, these data represent an assumed population and industrial expansion over time. The difference between the scenarios represents a move from present levels of treatment to an effluent standard for BOD at approximately half the current standard. The results of this better treatment policy are initially very good but decline with time. Looking at Table 2-5, we see that under AP, the stream goes below standards three times in 1975 and 1980 and five times in 1985. Under the stricter controls, the stream goes below standards only once in 1975. However, the advantage of this control strategy is mitigated by the projected future growth with the DO going below standards three times in 1980 and 1985. Thus, the investment in improved treatment will not pay off in this case unless accompanied by some growth controls or other methods of mitigating the impact of land use and population changes in the hypothesized future.

In a similar way, one can test the impact of a point source increase in discharge caused by the location of an individual plant and trace its impacts not only at the stream segment where it is located but also in the downstream segments. In each case, one can evaluate the impacts of the proposed changes in effluents, effluent standards, and plant locations.

REFERENCES

American Public Works Association (1969). *Water Pollution Aspects of Urban Runoff.* NTIS PB-215532.

Beck, M. B. (1978). "Modelling of Dissolved Oxygen in a Non-Tidal Stream," in *Mathematical Models in Water Pollution Control,* ed. by A. James. New York: John Wiley & Sons.

Clark, Leo J. et al. (1978). *A Water Quality Modelling Study of the Delaware Estuary.* Springfield, VA: NTISPB-282984.

Carey, G. W., L. Zobler, M. R. Greenberg and R. M. Hordon (1972). *Urbanization, Water*

Pollution and Public Policy. New Brunswick, NJ: Center for Urban Policy Research, Rutgers University. P. 150–158.

Environmental Law Reporter, 8 ELR 10010 (Jan. 1978).

Fair, Gordon Maskew, John Charles Geyer, and Daniel Alexander Okun (1971). *Elements of Water Supply and Wastewater Disposal.* New York: John Wiley & Sons.

Gordon, Steven I. and Richard K. Fromuth (1981). "A Point, Non-point Source Model of Dissolved Oxygen for the Great Miami River." *Journal of Environmental Systems,* Vol. 10, No. 3 (1980–81), p. 185–200.

Haith, D. A. (1976). "Land Use and Water Quality in New York, River," *Journal of the Environmental Engineering Division, Proceedings of the American Society of Civil Engineers,* Vol. 102, No. EE1 (Feb.), p. 1–28.

Hughto, Richard J. and Robert P. Schreiber (1982). "Microcomputer Water Quality Simulation Model," *Civil Engineering* (March).

SAS Institute Incorporated (1976). *A User's Guide to SAS 76,* by Anthony J. Barn, James H. Goodnight, John B. Sall, and Jane T. Helwig. Raleigh, NC: SAS Institute.

Streeter, H. W. and E. B. Phelps (1925). "A Study of the Pollution and Natural Purification of the Ohio River," Bulletin No. 146, U.S. Public Health Service.

Thomann, R. V. (1967). "Time Series Analysis of Water Quality Data," *Journal of the Sanitary Engineering Division, Proceedings of the American Society of Civil Engineers,* Vol. 93 (February), p. 1–23.

U.S. Army Corps of Engineers (1974). *Models and Methods Applicable to Corps of Engineers Urban Studies,* NTIS, AD-786516.

U.S. Environmental Protection Agency (1971). *Storm Water Management Model,* Water Pollution Control Research Series 11024-DO-C07/71. Washington, DC: U.S. Government Printing Office.

Wang, Multao, Lawrence, K. Wang, Jao-Fuan Kao, Ching-Gung Wen, and David Vielkind (1979). "Computer-Aided Stream Pollution Control and Management," Part I, *Journal of Environmental Management,* Vol. 9, p. 165–183.

3
Stormwater Runoff Models

INTRODUCTION

Stormwater is a continuing problem for both urban and rural areas. Not only does the excess runoff pose a potential flood hazard causing millions of dollars in damage each year, but it also produces environmental degradation in the form of erosion and sedimentation of lakes and streams and the transportation of water pollutants to these same water bodies. These water quality impacts were discussed in the previous chapter. The purpose of this chapter is to review in a more comprehensive way some of the many computerized stormwater models in use today.

It is first useful to note the circumstances under which stormwater modelling is being undertaken. Like some of the water quality efforts, stormwater analysis has both a regional and a local component. On a regional scale, agencies such as the Army Corps of Engineers, the Tennessee Valley Authority and various river basin and related commissions are concerned with the overall impact of land use decisions on the quantity and quality of water runoff in their areas of geographic interest. As urbanization proceeds, more of the precipitation that might have been intercepted by natural vegetation or seeped through the soil into the ground water becomes surface runoff entering lakes and streams. This transformation increases the severity and frequency of flooding and thus causes a major threat to life and property. The regional agencies are charged with the monitoring of trends in runoff and with making alternative policy recommendations aimed at offsetting the environmental risks. Such recommendations can include structural solutions (e.g. the building of flood control reservoirs) or non-structural solutions (e.g. controls over future land use to ameliorate the problems).

Aside from these regional scale studies, increasing use is being made of stormwater models at the local level. At this scale, both the flooding of rivers and lakes from their banks and the overflows of local storm sewers, culverts, and gutters are important. In heavily urbanizing areas, little attention has typically been given to the proper design of storm drainage systems or to the increases in runoff that occur during urbanization. Accordingly, stormwater becomes a pervasive local problem where small streams, pipes, and streets become flooded with increasing frequency. Local public officials are therefore faced with a rising number of angry calls to "do something about it."

As a result, there is an increasing tendency for the passage of local stormwater control ordinances. The exact form of these differs from community to community but most have the same basic, underlying premise—that stormwater poses a threat to the public health and welfare and therefore it should be regulated. The regulations generally impose standards for the storage of increases in stormwater, or of a certain volume of stormwater, on newly urbanizing sites rather than allowing the developer to pass extra runoff downsteam. An excellent discussion of some of the provisions of these laws can be found in Maloney et al. (1980). For now it is only important to note that in communities with such laws, any site development decision and subsequent plan must include an analysis of the stormwater impacts of the development and plans for ameliorating the consequences of those impacts in accordance with the provisions of the ordinance. Thus, there is an increasing need for the application of stormwater runoff models at this local scale.

Figure 3-1 illustrates the changes in runoff occurring as a result of urbanization. Here one sees the runoff hydrograph, a diagram showing the amount of stream discharge over time, for an example stream before and after urbanization. The scale on the y axis is discharge. Thus, the graph shows the changes in the water discharge over the time of a storm. The critical point is the discharge level at which a stream channel is filled to capacity. Discharge over this amount will result in a flood. After urbanization, the flood peak is higher and occurs more rapidly due to the changes in the hydrologic cycle discussed earlier. The goal of stormwater runoff controls is to avoid these dramatic changes through various design and management alternatives and thus to avoid the property damage and loss of life associated with flooding.

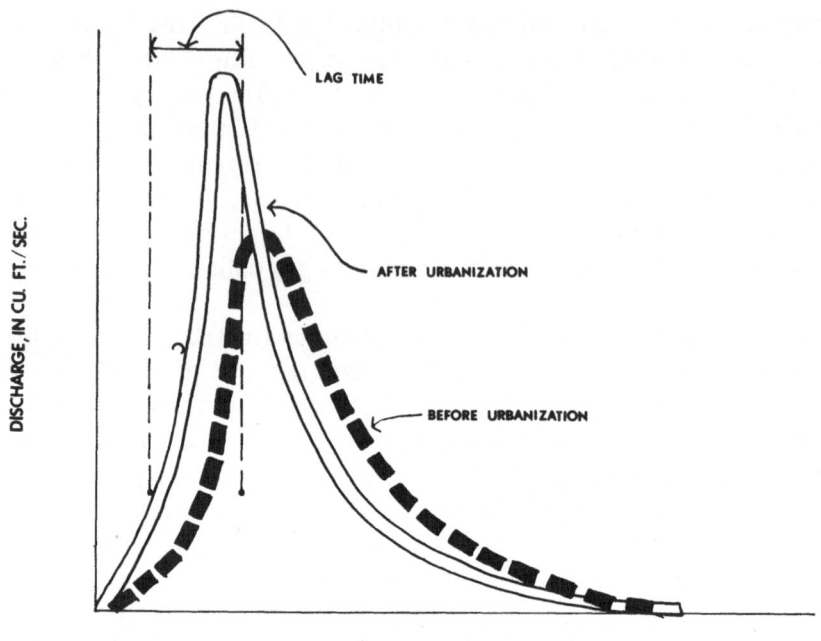

Figure 3-1. Runoff changes resulting from urbanization.

There are several steps associated with the analysis of stormwater impacts regardless of the scale. They can be summarized as follows:

1. Compilation of data on existing flooding problems and their frequency.
2. Determination of the relative contribution of various catchment areas to the runoff reaching drainage channels, both natural and manmade.
3. Determination of the amount of runoff which might come from new development, both with and without local controls and design changes.
4. Routing of the runoff water through the drainage system relative to the capacities of the drainage channels.
5. Determination of limitations in the capacities of drainage channels for handling stormwater runoff over time for various size storms and design alternatives.
6. Recommendations of policies aimed at ameliorating any stormwater runoff problems.

As we will see below, each of the above steps may require a separate modelling effort. At each step, there are choices in the manner of representation of the drainage system, each of which make different assumptions relative to the behavior of the system. Each also requires a different dataset and differing computer and manpower resources in order to be operationalized. We will first draw some theoretical distinctions between approaches to stormwater modelling and then present several existing, computerized models.

THEORETICAL CONSIDERATIONS: THE HYDROLOGIC CYCLE

The various theoretical approaches to the modelling of stormwater can best be illustrated by tracing the flow of water through the hydrologic cycle. Although water droplets are unthinking, begin by imagining yourself to be a thoughtful water droplet in a cloud trying to decide your own fate. The choices before you are to drop directly to the ground, to begin to drop and then evaporate, or to drop and be intercepted by vegetation, buildings, or other structures before reaching the ground.

Should you reach the ground, several things could happen. You could be absorbed by the surface soil, there to remain in storage or to begin flowing through the soil toward the ground water or outward toward an open body of water. Instead of being absorbed by the soil, you could be stored on the surface of the soil in small surface depressions, there to either evaporate or wait until the depression fills enough to allow you to flow along the surface. If enough water drops have already filled the depressions and exceeded the capacity of the soil to absorb them, you could flow over the surface of the soil as runoff and move directly toward the nearest surface body of water. Obviously, water droplets are not given to prescient choices of behavior. However, if one thinks through this analogy, one can gain some insights into the hydrologic cycle and therefore into modelling the fate of water falling as precipitation.

Put in more traditional terms, one says that inflow (of water to the hydrologic system) must be exactly equal to outflow plus or minus storage. *Inflow* is always expressed in terms of precipitation while *outflow* can be one of the many pathways discussed above. How much outflow takes each path is dependent upon a number of conditions.

Interception of precipitation near the surface is dependent upon the nature and extent of surface cover. Trees are most effective with coniferous forest cover allowing the greatest amount of interception because the finer leaves allow for a higher rate of evaporation and prevent the formation of large droplets which fall to the ground. Surface absorption is dependent upon the soil surface and subsurface conditions. Organic litter will have a high water absorption capacity. Sandy soils will allow for the faster infiltration of water while heavier, clay soils will impede this infiltration. Soils which have been compacted by construction or farm equipment or animals will have a reduced infiltration capacity.

Similarly, the amount of surface storage is related to the soil conditions and topography. Coarser soils will tend to have more surface storage. High slopes will not allow for the storage of much water before runoff begins. If the natural soils are covered with pavement and buildings, little or no surface storage or infiltration will occur.

All of these events will be influenced in addition by the nature and timing of the precipitation. If no precipitation has occurred for some time, the soil's water holding capacity will be greater. A storm of higher intensity and shorter duration will allow less time for surface absorption than a less intense, longer storm with the same total rainfall.

A number of models have been proposed as representations of these complex processes. They vary in the complexity of their representation and therefore in the accuracy with which they portray the runoff of stormwater. At the first level of generalization, most models simplify the nature of the storm events relative to two variables, *intensity* and *duration*. Storm records are analyzed and represented in terms of their probability of occurrence with respect to these variables. Almost all stormwater runoff models use a representation such as this to simulate the response of given areas to a range of potential events. There is evidence that other storm factors such as storm path, shape, and sequence will also impact the runoff patterns (Linsley et al., 1975; Ward, 1967).

The next stage of the runoff process involves the response of the basin soils, vegetative, and manmade cover to specific storm conditions. Here, models differ widely in their level of generalization.

Horton's Theory

One of the first theoretical treatments of the rainfall-runoff relationship was proposed by Horton (1945). Horton's theory involves both the erosion and runoff processes. He noted that rainfall falling on the bare earth will infiltrate into the ground at a rate related to soil texture and structure, vegetative cover, soil moisture content, and surface condition. Initially, this infiltration capacity f_p, the maximum sustained rate of infiltration, will be very rapid and proceed at a rate f_o but then decline to a constant value f_c. During long rainstorms, this minimum infiltration value is in effect and produces a relatively constant level of runoff. Over a period t of rain the following formula applies:

$$f_p = f_c + (f_o - f_c) \, e^{-kt}$$

where k is a constant and e is the base of the Naperian logarithm. In this model, k is thought to represent variations in soil and surface conditions.

In practice, the formula is difficult to apply because there are few data on the infiltration constants and variables. Thus, several approximations have been devised. The simplest case and the one used most historically is the so-called *rational method*. This model is represented by the following equation:

$$Q = CiA$$

where

Q = the peak discharge in cubic feet per second;
C = a coefficient of runoff;
i = the rainfall intensity in inches per hour for a duration equal to the time of concentration t;
A = the drainage area in acres.

Theoretically, the time of concentration is the time it takes for a particle of water in the farthest part of the drainage area to reach the outlet. A number of representations of this time have been derived and presented in both mathematical and graphical form (see Overton and Meadows, 1976; ULI, 1975). In this equation, one can see that the basin response is represented in terms of one simple runoff coefficient. This varies with respect to land use and soil type as shown in Table 3-1. No representation is made of the impacts of antecedent storms, surface storage changes, or other factors affecting the rate of runoff (slope, vegetation, and so on.) Weighted averages based on basin subareas can easily be used to calculate the runoff using this equation. Because of its simplicity relative to data requirements and ease of calculation, this model is frequently used for the first estimation of runoff amounts. This model represents only an empirical approximation of runoff and not any theoretical treatment of the process.

Other representations based more closely on Horton's work take advantage of other empirical approximations to his ideas concerning runoff. Several use multiple criteria to decide on the amount of infiltration which might occur and thus include calculations relating to antecedent moisture conditions (whether another storm has occurred in the recent past), soil type, vegetative cover, storm intensity, and so forth. Several of these factors are included in the models that will be discussed later in the chapter.

Unit Hydrograph Method

Another approach to the modelling of stormwater runoff is that of the *unit hydrograph method*. This method, originally proposed by Sherman (1932), is based on the principle that the hydrograph of a stream reflects the physical characteristics of the basin and the character of the rainfall which has occurred. As such, similar rainfalls with comparable antecedent moisture conditions will produce hydrographs of comparable shapes. Further, the method defines a typical or unit hydrograph corresponding to one set of conditions that reflects the basin characteristics. Once defined, this unit hydrograph can be used to estimate the runoff and stream hydrograph for storms of any intensity or duration, in proportion to the number and timing of unit hydrographs. In other words, unit hydrogaphs are superimposed to reflect whether a storm of multiple unit intensity or longer time period relative to the unit hydrograph might occur.

Looking back at Figure 3-1, one might imagine that the hydrograph labelled "before urbanization" represents the shape or response of that basin at that time. This could be the unit hydrograph

Table 3-1. Runoff Coefficients

Description of Area	Runoff Coefficients
Business	0.70 to 0.95
Downtown	0.50 to 0.70
Neighborhood	
Residential	
Single family	0.30 to 0.50
Multi-units, detached	0.40 to 0.60
Multi-units, attached	0.60 to 0.75
Residential (suburban)	0.25 to 0.40
Apartment	0.50 to 0.70
Industrial	
Light	0.50 to 0.80
Heavy	0.60 to 0.90
Parks, cemeteries	0.10 to 0.25
Railroad yard	0.20 to 0.35
Unimproved	0.10 to 0.30

Character of Surface	Runoff Coefficients
Pavement	
Asphalt or concrete	0.70 to 0.95
Brick	0.70 to 0.85
Roofs	0.70 to 0.95
Lawns, sandy soil	
Flat, 2 percent	0.05 to 0.10
Average, 2 to 7 percent	0.10 to 0.15
Steep, 7 percent or more	0.15 to 0.20
Lawns, heavy soil	
Flat, 2 percent	0.13 to 0.17
Average, 2 to 7 percent	0.18 to 0.22
Steep, 7 percent or more	0.25 to 0.35

The coefficients in these two tabulations are only applicable for storms of five to ten year return frequencies, and were originally developed when many streets were uncurbed and drainage was conveyed in roadside swales.

For recurrence intervals longer than ten years, the indicated runoff coefficients should be increased assuming that nearly all of the rainfall in excess of that expected from the ten year recurrence interval rainfall will become runoff and should be accommodated by an increased runoff coefficient.

The runoff coefficients indicated for different soil conditions reflect runoff behavior shortly after initial construction. With the passage of time, the runoff behavior of sandy soil areas will tend to approach that of heavy soil areas. If the designer's interest is long-term, the reduced response indicated for sandy soil areas should be disregarded.

Source: *Design and Construction of Sanitary and Storm Sewers.* ASCE Manual of Practice No. 37, 1970. Notes revised by D. Earl Jones, Jr. Also in ULI, ASCE, NAHB, 1975.

for this basin. If the rainfall causing that response is known, then differences in rainfall for other storms can be apportioned to yield a new composite hydrograph using this "unit" as the building block. The method then yields a predicted hydrograph for the simulated conditions. If the results show the stream to go over capacity, a simulated flood has occurred, something we wish to avoid in reality. The method requires that assumptions be made about the relationships between infiltration capacity and rainfall intensity and that storms over the basin are uniform in distribution. These assumptions can lead to significant errors in the method (see Ward, 1967; Henderson, 1963, cited in Ward).

Once one has calculated the amount of runoff that may be produced in a drainage area, this discharge must be routed through the basin. This process, called *flood routing,* has been subjected to a number of different modelling approaches. One major distinction can be made here between models that are mostly applied to natural channels versus those which are applied to urban gutter and pipe stormwater drainage systems. Obviously, combinations are also possible. Some methods, such as the unit hydrograph method, simply predict the shape of the hydrograph of natural streams irrespective of direct consideration of basin physical characteristics. Others attempt to approximate the behavior of the flood as it moves downstream. The *Manning equation* is frequently used to estimate the capacity of channels, pipes, and streets to accommodate the flow of water. In its most basic form

$$V = \frac{1.486}{n} R^{2/3} S^{1/2}$$

where

    V = velocity in feet per second;
    R = the hydraulic mean radius in feet calculated as cross-sectional area divided by the wetted perimeter (this is the bed and sides of the channel which are "wetted" when water flows in it);
    S = the rate of loss of head per foot of channel (channel slope);
    n = a roughness coefficient.

The major problem with the equation lies in the estimation of the roughness coefficients. For purposes of calculating flow, velocity is converted to flow based on its relationship to cross-sectional area:

$$Q = VA$$

then

$$q = \frac{1.486}{n} AR^{2/3}S^{1/2}$$

The hydraulic capacity of streets can be calculated using a different form of this equation:

$$Q = 0.56 \frac{Z}{n} d^{8/3}S^{1/2}$$

where

Q = discharge in cfs;
Z = $1/S_x$ where S_x is the cross-slope of the pavement;
d = depth of water in feet at face of curb;
S = longitudinal grade of street;
n = Manning's roughness coefficient (ULI, 1975, p. 38).

The results of any Manning calculations must be compared to the capacity of the structure or channel to transmit water. If a constriction occurs, water will back up in the upstream direction. If not, water will continue to move downstream, collecting in additional channels and pipes. The resultant water movement can be thought of as a wave moving in either the upstream or downstream direction. Here, waves must be thought of in a special sense, ". . . a temporal spatially propagated change in the water surface." (Yevjevich, 1975, p.2) The next analytical step must be to determine the size and location of this wave in order to approximate the location of peak flow and the equivalent stream or pipe hydrograph.

A number of flood routing methods have been derived to model this wave movement. Here, we will discuss only one since it is the method used in two of the models that we will review. This is called the *kinematic wave,* a type of wave that makes some simplifying assumptions in order to allow for easier calculation of the flood movement. In particular, a kinematic wave utilizes two governing equations: a one-dimensional unsteady flow continuity equation and a steady, uniform flow momentum equation. Discharge then becomes a function of depth alone, which in turn is a function of rainfall excess, surface roughness, length and slope of the catchment area.

There are a number of different representations of this model, each of which makes different assumptions. The reader is referred to other works for a description of the theory behind this model (see Yev-jevich, 1975; Miller and Cunge, 1975; Overton and Meadows, 1976). The assumptions of the specific computer models will be discussed below. The importance of the kinematic wave is in yielding a stream or pipe hydrograph over space and time that can be compared with capacities and that describes the stormwater impacts of precipitation on a given basin.

MODEL DESCRIPTIONS

Introduction

A very large number of stormwater runoff models have been computerized. Only three have been selected for discussion here. These are the Soil Conservation Service method of runoff and flood routing (TR20), the Storm Water Management Model (SWMM) produced for the U.S. Environmental Protection Agency, and the STORM model of the U.S. Army Corps of Engineers. These models were chosen because of their widespread use, their differences in resource requirements, and their differences in theoretical approach. They are representative of the choices available to the analyst who does not wish to use an entirely empirical approach such as the rational method in studying stormwater problems. Each of these models is discussed below, with a comparison and some examples following.

The Soil Conservation Service Hydrology Program

The U.S. Department of Agriculture, Soil Conservation Service, has developed a series of procedures for calculating the stormwater impacts of land use changes and the construction of various types of structures aimed at controlling runoff. The initial work in this area was in the form of a set of numeric and graphical procedures for performing calculations by hand. This information (see USDA, SCS, 1972) provides a detailed description about the methods and their solution. In parallel with this document, the SCS has developed a computer program that performs the major calculations for runoff and flood routing available in the other document (USDA, SCS, 1965). We will first describe the nature of the methods used in the calculations and then review the computer program capabilities.

The SCS methods are somewhat unique in providing fairly detailed adjustments of runoff data based on a number of surface soil, vegetation and cover, and condition variables. The determination begins with the classification of the soils of the area into the proper hydrologic soil group. This is a designation which represents the infiltration and transmission response of the bare soil to standard rainfall conditions. Hydrologic soil groups A through D represent the soils with the highest and lowest infiltration rates respectively. Tables given by SCS relate soils series names, available from detailed soil survey documents, to the corresponding hydrologic groups.

The next variable of significance in the SCS method is called the *hydrologic soil-cover complex.* This variable incorporates the hydrologic soil group and two other variables, the land use, and the hydrologic condition of a parcel of land For each combination of these variables, the method defines a curve number, or CN, which represents the rainfall-runoff relationship for that land parcel; see Table 3-2.

Direct runoff from storm rainfall is the next item to be determined. Here, the SCS method assumes a relatively constant initial abstraction of rainfall, i.e. the surface storage, that occurs prior to runoff. The model also defines three antecedent moisture conditions:(average, wet, and dry. Given these items and the CN number, one then determines the direct runoff corresponding to a given rainfall amount. This concludes the first portion of the runoff modelling.

In the second part of the model, runoff is routed through a stream system. The stream is represented schematically by a series of stream reaches, cross-sections, and structures for which data are available or can be estimated. Figure 3-2 shows a simple example from the SCS manual. The routing of the flood requires the sequential calculation of inflow and outflow hydrographs for each reach and structure.

Table 3-2. Runoff Curve Numbers for Hydrologic Soil Cover Complexes
(Antecedent moisture condition II, and $I_a = 0.2\ S$)

Land Use	Cover Treatment or Practice	Hydrologic Condition	Hydrologic Soil Group A	B	C	D
Fallow	Straight row	—	77	86	91	94
Row crops	"	Poor	72	81	88	91
	"	Good	67	78	85	89
	Contoured	Poor	70	79	84	88
	"	Good	65	75	82	86
	"and terraced	Poor	66	74	80	82
	" " "	Good	62	71	78	81
Small grain	Straight row	Poor	65	76	84	88
		Good	63	75	83	87
	Contoured	Poor	63	74	82	85
		Good	61	73	81	84
	"and terraced	Poor	61	72	79	82
		Good	59	70	78	81
Close-seeded legumes[1] or rotation meadow	Straight row	Poor	66	77	85	89
	" "	Good	58	72	81	85
	Contoured	Poor	64	75	83	85
	"	Good	55	69	78	83
	"and terraced	Poor	63	73	80	83
	"and terraced	Good	51	67	76	80
Pasture or range		Poor	68	79	86	89
		Fair	49	69	79	84
		Good	39	61	74	80
	Contoured	Poor	47	67	81	88
	"	Fair	25	59	75	83
	"	Good	6	35	70	79
Meadow		Good	30	58	71	78
Woods		Poor	45	66	77	83
		Fair	36	60	73	79
		Good	25	55	70	77
Farmsteads		—	59	74	82	86
Roads (dirt)[2]		—	72	82	87	89
(hard surface)[2]		—	74	84	90	92

[1]Close-drilled or broadcast.
[2]Including right-of-way.
Source: USDA, SCS, 1972

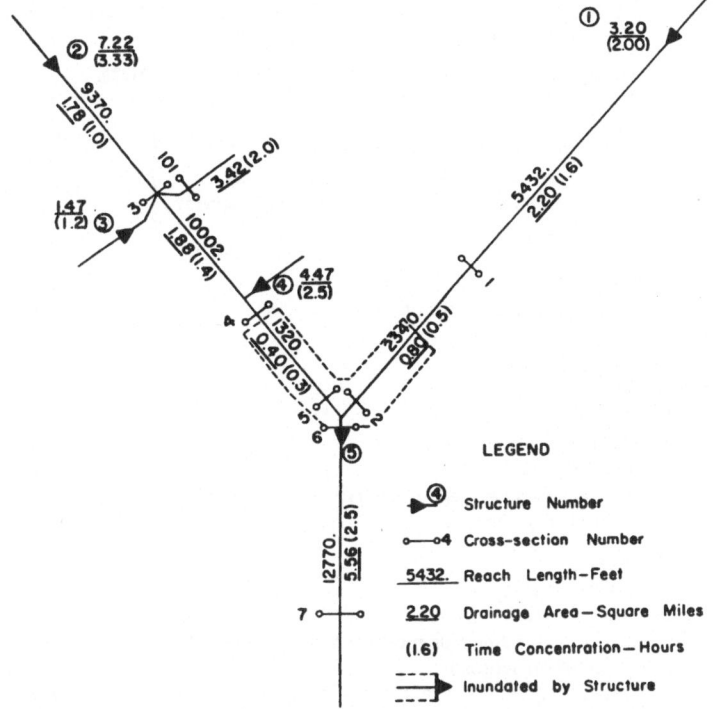

Figure 3-2. Routing runoff through a stream system. (Source: USDA, SCS, 1965.)

Although several routing methods are available in the hand calculation version of the method (unit hydrograph, mass curve, storage indication, convex method), the computer program uses only one method each for reservoir and channel routing (the storage-indication and convex methods respectively) combined with a unit hydrograph determination of hydrograph shape. The output of the process is both the hydrograph for each cross-section and structure and the peak flow at each location.

Unfortunately, the original documentation for the computer model was written for internal consumption only and thus is very cryptic and difficult to understand. One must refer to two other documents in order to fully understanding the nature of the computations being made (USDA, SCS, 1975 (tech. rel. 55); USDA, SCS, 1972). The program is written in Fortran and is readily available from the SCS state office in most states or the National Technical Information Service. Thus it will run on most mini and mainframe

Table 3-3. Computer Program for Project Formulation Hydrology (Referred to as TR-20) Input Requirements and Program Outputs

Inputs	• Drainage area
	• Hydrologic soil class derived from soil type
	• Antecedent moisture classification and % land use
	• SCS curve number derived from soil type, % impervious cover. and % land use
	• Travel times for flow through sub-basins
	• Times of concentration for sub-basins
	• Rainfall depth and distribution
	• Frequency information on the rainfall
	• Baseflow data (optional)
	• Basin length
	• Basin slope
	• Channel cross-sections
	• Structure geometry (if any)
	• Stage-discharge and storage-discharge* relationships for channels and structures
Outputs	• Peak flood flow
	• Flood hydrograph
	• Flood water surface elevations
	• Outputs can be for single or multiple basins (up to 200 reaches and 99 structures)

The first four inputs are grouped together with the annotation: Use to calculate CN, exogenous to computer

*For multiple basins only.

computers. Once one deciphers the documentation, the model is relatively easy to run. However, it is also in a rather inconvenient, batch processing format, which makes data compilation and multiple run analysis very cumbersome and time consuming to perform.

Table 3-3 shows the model input requirements and outputs for the TR-20 model. One of the major inconveniences of the model is that all of the calculations for the CN number must be performed exogenously to the computer model and fed in as input data. This means that the analyst must compile data on land use, soils, and make hydrologic-condition assumptions. Then. for each stream reach. a weighted CN number must be determined based on these data and input to the computer model. Antecedent moisture is the only variable relating to surface conditions that the model will manipulate automatically. The model can use either an idealized unit hydrograph or any actual storm as input. Alternative combinations of land use, structure placement and design, and channel improvements can be tested using this model. Thus, most of the major policy decisions discussed in the introduction can be made using this model.

Information on the accuracy of the model is rather scanty. It is highly touted by the SCS as a rather simple and straightforward method for the calculation of stormwater impacts particularly suited to analysis of rural and semiurban or urbanizing watersheds. It does not, however, provide enough detail to enable calculations to be performed for heavily urbanized watersheds. Some authors seem to indicate that the CN number of runoff calculation is acceptably accurate (Overton and Meadows, 1976) but then go on to deride the method of routing calculations. "Although the prediction technique of the SCS is entirely specified, little information concerning its reliability has been reported in the open literature." (Overton and Meadows, 1976, p. 37) Espey and Winslow (1975) generally criticize methods based on the unit hydrograph as being unrealistically linear while the runoff process is non-linear. Overall, despite any criticisms, the methods used by the SCS seem to be widely accepted and used in practice. As we will see in the following examples, the model can be used effectively in making planning decisions.

The Storage, Treatment, Overflow, Runoff Model "STORM"

The STORM model is the answer of the U.S. Army Corps of Engineers to other models developed for the simulation of stormwater impacts. It is written in Fortran with versions readily available for CDC, IBM, UNIVAC and certain other (not named in the documentation) computer systems. The model is much more comprehensive in scope than the SCS model, encompassing both quality and quantity calculations. The model's overall capabilities are illustrated by Figure 3-3. The model can be used to track the interaction of rainfall-snowmelt, runoff, dry weather flow, pollutant accumulation and washoff, land surface erosion, treatment rates, and detention reservoir storage. (U.S. Army Corps of Engineers, 1977, p. 2)

In actual practice, the numeric methods used in the model overlap substantially with the SCS model and with other approaches. Runoff quantity can be calculated using a coefficient method or the SCS curve number method. The coefficient method accounts for depression storage and any water diversion and then calculates runoff using a derivation of the rational method. The SCS curve number method has been discussed. STORM has no flood routing methodology to help produce an output hydrograph. It depends either on the unit

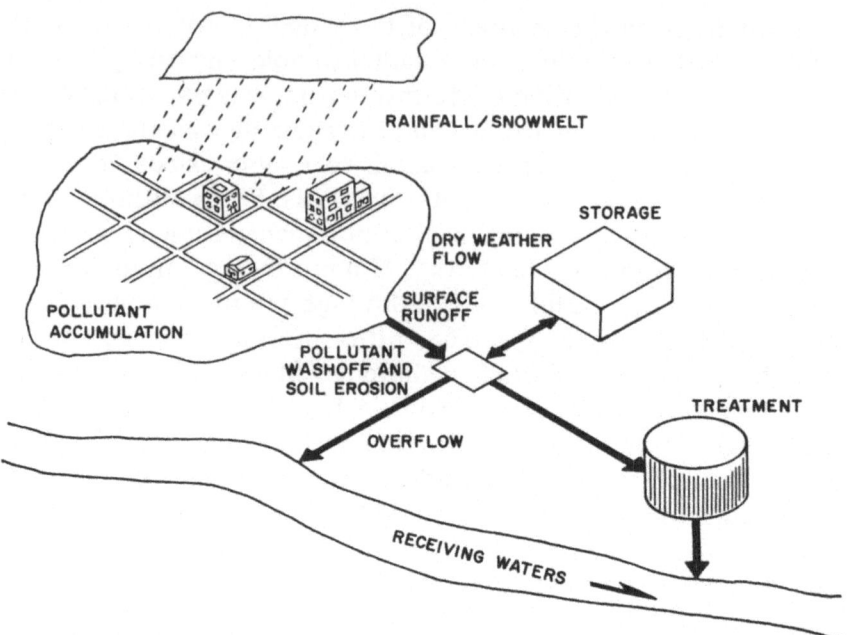

Figure 3-3. Major processes modelled by storm. (Source: U.S. Army Corps of Engineers, 1977.)

hydrograph or the rational formula to produce an estimate of peak flow and assumes the hydrograph shape to be constant. STORM also accounts for evapotranspiration, infiltration, and percolation for times when there is no precipitation. Baseflow, which can be computed using the TR-20 model, cannot be calculated using STORM. The soil moisture capacity equation used in STORM is

$$S_t = S_{t-1} - IN*\Delta t + A*EV*\Delta t + B*MP*\Delta t$$

where

$A = 0.7 ((SM - S_{t-1})/SM)^v$;
$B = ((SM - S_{t-1})/SM)^p$;
S = soil moisture capacity for storage of water in inches (mm);
IN = maximum infiltration rate from initial abstraction in inches/hour (mm/hour);
EV = pan evaporation rate in inches/hour (mm/hour);
MP = maximum soil percolation rate in inches/hour (mm/hour);

SM = maximum soil moisture capacity for storage of water in
    inches (mm);

  t = time;

Δt = 1 hour;

  v = exponent regulating evapotranspiration;

  p = exponent regulating percolation.

The exponents v and p must be calibrated by comparing model
results with observed data. Values have been found to be between 1.0
and 5.0.

The calculations for dry weather flow and quality of runoff are
made using coefficients similar to those (and in some cases the same
as) in the Storm Water Management Model. (These quality calculations were discussed in Chapter 2.) Treatment options can also be
evaluated using this model.

The component of the STORM model which makes it different
from both the TR-20 and SWMM models is in the calculation of soil
loss from a study area or subbasin. Here the model uses the method
of the Universal Soil Loss Equation (USDA, 1965 Ag. Hndbk. 282)
to estimate the amount of soil loss that might occur in an area based
on soil type, slope amount and length, cropping practices, conservation practices, and rainfall intensity. A sediment delivery ratio is used
to estimate the total sediment which might reach local streams. This
is one of the few models which allows one to make these calculations
using the computer.

Table 3-4 illustrates the input requirements and outputs of the
STORM model. Although the data requirements are substantial for
using all of the modelling options, the stormwater portion of the
model requires a minimum of input data, particularly if one uses the
rational method type of calculation. This lack of sophistication is
problematic if one has a relatively complex basin. Thus, the model
should generally be used only for small, uncomplicated stream
situations. For such basins, the initial estimation of the impacts of
land use changes and stormwater controls is rather straightforward
using STORM, making it well suited to such calculations.

The criticisms that have already been aimed at the rational
method, the SCS and unit hydrograph method, and the stormwater
quality calculation methods all apply to the STORM model. The

Table 3-4. Storage, Treatment, Overflow, and Runoff Model
STORM
Input Requirements and Outputs

Input	• Hourly precipitation data
	• Surface depression storage
	• Runoff coefficients for pervious and impervious portions of urban areas and for various portions of rural areas
	• Potential evaporation for each month for urban and non-urban portions of the watershed
	• Mean daily or maximum/minimum air temperatures for the period coinciding with the precipitation record
	• % land use and % impervious cover associated with each land use
	• Street and gutter density and street sweeping interval associated with each land use (up to 20 land use categories may be included in the model)
	• Default values for water quality parameters are supplied if no data is available
	• SCS curve numbers if the SCS runoff method is used
	• Input unit hydrograph if rational method is not used
	• Soils data, slope, and management factors for Universal Soil Loss Equation
Output	• Duration and amount of rainfall
	• The amount of runoff
	• Quantity and number of hours of treatment
	• Duration of storage
	• Ages of storage
	• Overflow to receiving water
	• Annual summary tables
	• Runoff quality
	• Overflow quality
	• Detailed hourly listings of rainfall, runoff, amount of pollutant washed off during each hour of the event, and pollutant concentration

novice user must therefore be cautious in the interpretation of model output due to the errors that are embedded in these methods.

The Storm Water Management Model (SWMM)

The Storm Water Management Model (SWMM) was originally formulated for the USEPA by Metcalf and Eddy, the University of Florida, and Water Resources Engineers (USEPA, July, 1971). More recent versions of the model have been compiled by the University of Florida. The purpose of the model is to "assist administrators and engineers in the planning, evaluation, and management of overflow abatement alternatives." (USEPA, July, 1971, p. iii) The model is one of the most complex models of stormwater runoff and quality in widespread use today. The model has the following major capabilities:

1. Conversion of any rainfall hyetograph (a graph showing the time distribution of rainfall) or hyetographs into a runoff hydrograph for each watershed that is modeled.
2. Production of pollutographs, runoff water quality graphs, based on land use, management, rainfall, and other data.
3. Computation of hydrographs and pollutographs for dry weather flow.
4. Flood routing through multiple channels and pipes.
5. Simulation of the impacts of optional water storage facilities.
6. Computation of stormwater treatment impacts using an array of treatment technologies.
7. Computation of receiving water quantity and quality impacts.
8. Computation of capital and operating and maintenance costs for storage and treatment options.

The model is written in Fortran, allowing for its application on many different mainframe computer systems. A large machine is needed to meet the minimum core requirement of this model of 350K bytes. The model is relatively complex in structure, consisting of five major subprograms, each performing a particular set of computational tasks. Figures 3-4 and 3-5 represent the conceptual and programming information flows respectively to the model. (Figure 2-4 also shows the conceptual information flows.) From these representations one can see that this model constitutes a very comprehensive approach to the analysis of stormwater runoff quality and quantity. The model also allows the handling of a large amount of data enabling the user to represent the flow of water and pollutants through the system at a relatively high level of detail. The major drawback to this detail is that it implies a correspondingly high cost, estimated at between "ten thousand to several hundred thousand dollars per city, depending on the size, complexity, and depth of investigations." (USEPA, July, 1971, p. 37) In order to illustrate this complexity, each of the major program blocks will be reviewed relative to input requirements and model outputs, and include a discussion of model assumptions and characteristics.

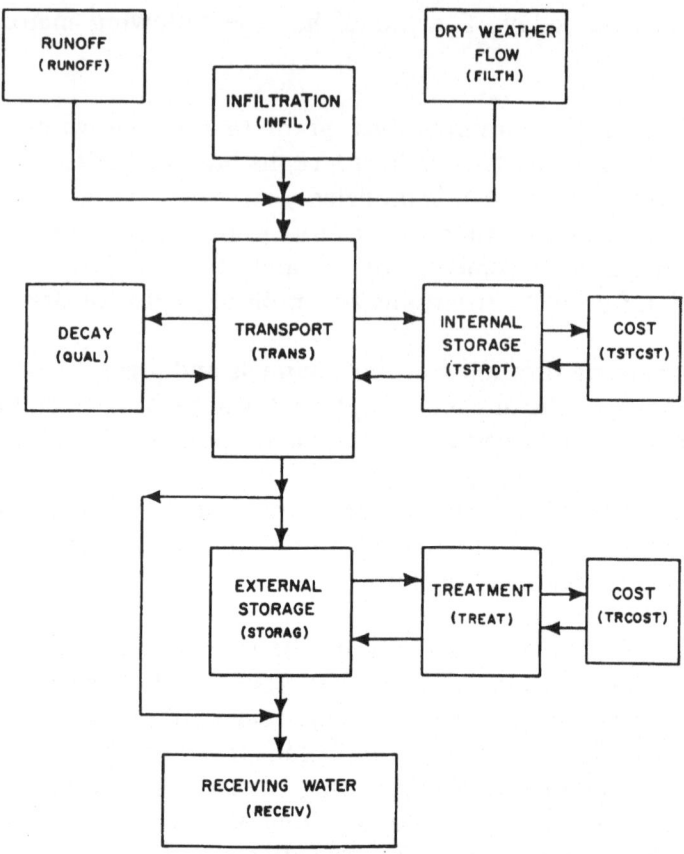

Figure 3-4. Overview of model structure, SWMM. Subroutine names are shown in parentheses. (Source: USEPA, 1971.)

The first segment of the SWMM model is called the Executive Block. The major purpose of this subprogram is to control the order of execution of the other subroutines and to handle the transfer, in and out, of various data files, and graphical and printed output. Input data for this program segment consists of titling information, general stream flow information, design flow rates and available sewer capacity, storm data for each storm being studied, and input and output file assignments. The executive block is also used to call the other subroutines in the form of simple name control cards. Up to 105 different parameters may be assigned or manipulated in this segment of the program. Only a few of these will be used at one

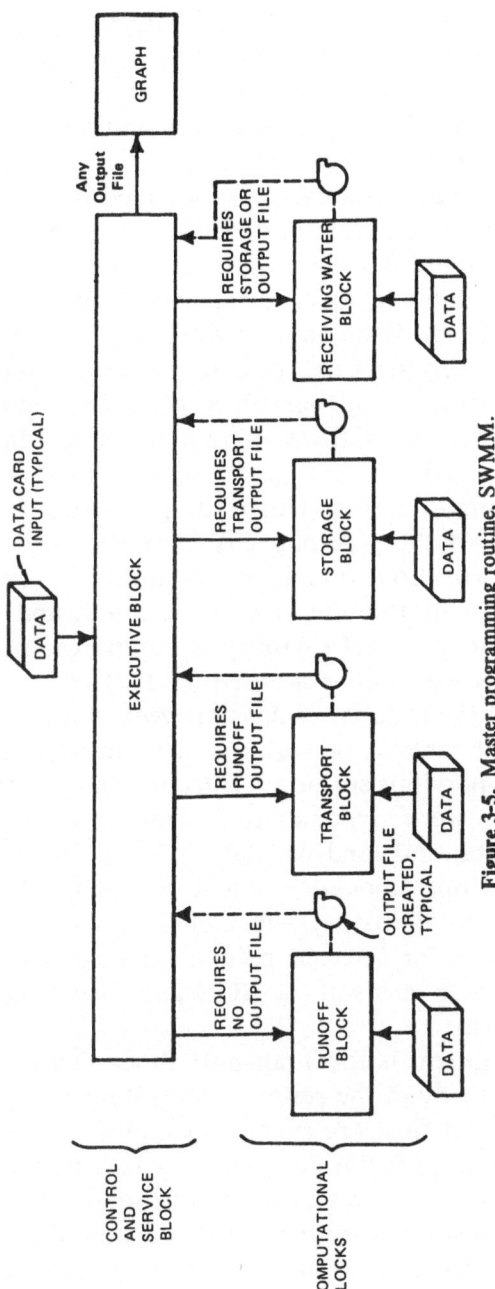

Figure 3-5. Master programming routine, SWMM.

time with many being left to assume default values. However, the sheer number of potential options illustrates the complexity of this model.

The next segment of the SWMM is called the Runoff Block. The purpose of this subprogram is to simulate the quantity and quality of surface runoff. The drainage basin can be subdivided into up to 100 subcatchment areas. These can drain in turn into a maximum of 100 pipes or gutters. These pipes or gutters represent the inlet points for the Transport Block. Subcatchment elements of overland flow receive rainfall in the form of an input hyetograph, use Horton's formula to account for infiltration, calculate surface storage or retention, and pass an output hydrograph to a gutter or pipe. The gutters or pipes allow for the outflow of rainfall using Manning's equation. The flow is calculated one gutter or pipe at a time and routed in the downstream direction. This routing is dependent upon kinematic wave theory. If the pipe capacity is exceeded, it will be allowed to surcharge—to hold extra water—until the excess is drained off. The output will note the location, amount, and time of all such occurrences. Once runoff quantity is computed, pollutant removal is computed for each subarea based on a set of coefficients relating to land use and the number of days between storms.

About 60 input data items may be required including data on each subcatchment area—impervious surface, infiltration rate; each gutter or pipe's number, size, slope; calculation instructions; catchbasin data; street cleaning data; and land use data. Almost 300 variables are used in the calculations associated with this submodel. Model output here shows the rainfall history, subcatchment and gutter-pipe data, output hydrographs for selected points, graphed hyetograph and output water quality in terms of the BOD and suspended solids removed during the storm.

The next program segment is the Transport Block. This controls the routing of the flow through the sewer system. Both the quantity and quality components of flow are routed through sewer elements such as conduits, manholes, lift stations, overflow structures, and so forth. Infiltration both during the storm and during dry weather is also computed in this portion of the model. This portion of the model requires very detailed data concerning the sewer system including locations, shapes, and sizes of all sewer system elements, unit costs, infiltration rates, sewer flow variations by subareas, capacities of storage elements, and many calculation parameters. About 150 data elements and parameters can be used for input data with several hundred variables used for model computations.

The Storage Block is an optional subprogram which models the effects of storage devices and treatment processes on water quantity and quality. Input data requirements include treatment method and parameters, storage structure and parameters, and treatment costs per unit treated.

The final segment of the program, the Receiving Block, simulates the quantity and quality of water in natural bodies of water. Data requirements include hydraulic data, information for tidal streams, input hydrograph, and quality data for each channel section. Many of these data can be transferred as part of the output from the Transport or Storage Block. The output includes an hydrograph and concentrations of major pollutant constituents.

Verification and testing of the SWMM has been carried out for several test sites. (USEPA, 1971, vol. II) Verification results seem to show that the model generally approximates the real data for most of the calculations carried out using the model. However, in most instances the results are not reported in statistical terms. The possibility of some errors become obvious as one reviews the verification results. Figures 3-6 and 3-7 show some examples for a test site in Cincinnati. Water quantity results are much more accurate than those for quality. This is expected given the nature of the models applied in both circumstances. Overall, the model appears to be at least as reliable as other approaches. Its major advantage lies in the level of detail that can be represented, particularly for heavily urbanized stormwater systems. This virtue also provides a major disadvantage, requiring a tremendous amount of resources to accumulate the data necessary to run the model.

Problems in the Models

As the previous discussion alludes, the output for each of the stormwater runoff models is voluminous. For this reason, we will only show portions of the output in illustrating the types of analyses which can be carried out using the models.

Let us begin with an example using the SCS, TR-20 model. The scenario we will draw upon is one in which a major land use change is

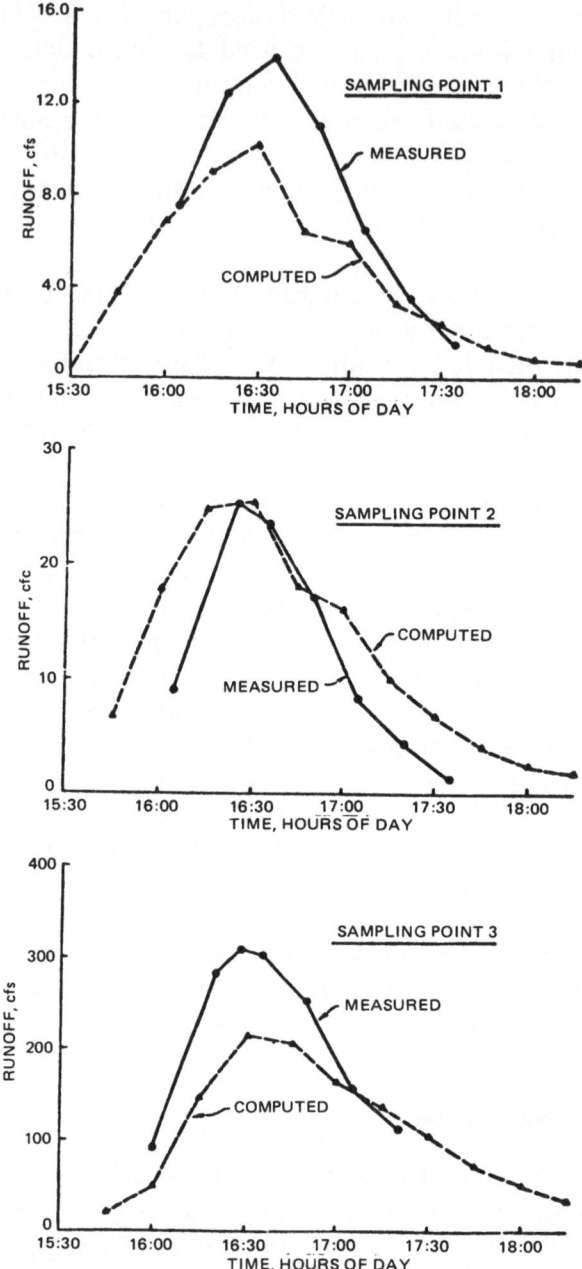

Figure 3-6. Comparisons between measured and computed hydrographs for storm of April 1, 1970, Cincinnati, Ohio. (Source: USEPA, 1971, Vol. II.)

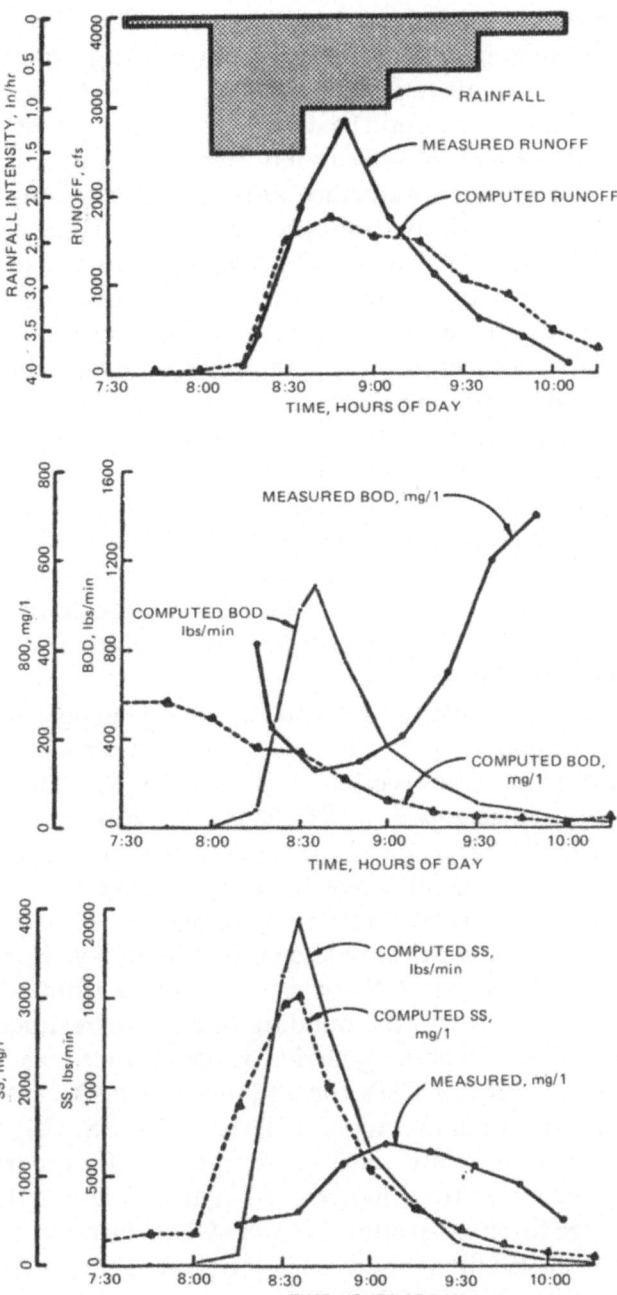

Figure 3-7. Combined sewer overflow results for storm of May 12, 1970, Cincinnati, Ohio (sampling point 3). (Source: USEPA, 1971, Vol. II.)

being proposed in a relatively small area of a watershed. In the base case, our urbanizing area still has a relatively large amount of open area, giving it an average CN of 70.0. Using the standard unit hydrograph, and data for the ten year storm (the storm with a recurrence probability of once in ten years), the model simulates the runoff coming from the watershed area and gives as output the peak discharge and runoff hydrograph over time. These outputs are labelled on Figure 3-8. The output hydrograph must now be computed for the next cross-section downstream, cross-section 3, and added to the runoff from cross-section 2. This takes four steps in the program. First, the output hydrograph from cross-section 2 is saved using a SAVMOV statement. Next, the computer is told to route the discharge through cross-section 3 with the REACH statement. Then, the RUNOFF statement is used again, this time to calculate the runoff originating from the watershed area corresponding to cross-section 3. These steps are shown in the first four sections of Figure 3-8. Finally, the output hydrographs from the two runoff steps and one routing step are added together to show the total stormwater being routed through cross-section 3. This calculation is shown in the ADDHYD part of Figure 3-8.

It is the initial runoff figures and the final hydrograph which are interpreted for making policy decisions. In Figure 3-8 we see that the peak runoff from cross-section 2 is 218.128 cfs. The total peak discharge at cross-section 3 is 385.356 cfs which translates to a peak elevation of 1119.16 feet. From the input data for cross-section 3 (not shown here), the bankfull stage for cross-section 3 is 1119.50 feet. Thus, this stream is barely below flood stage for this storm. We can compare these data with those shown in Figure 3-9. Here, the only alteration made is in the CN for cross-section 2 from 70.0 to 90.0. This could happen due to the building of a major regional shopping area, for example. To actually arrive at the new CN, we would take the weighted average of CN's for all land uses in the subcatchment area using information in Technical Release 55 (USSCS, 1975).

Given such a dramatic change in the runoff, the model also produces some interesting changes. In Figure 3-9 we see that the new peak discharge for cross-section 2 is 588.696, an increase of about 370 cfs. This translates into an increase in the peak runoff at cross-section 3 of 187 cfs (from 385 to 572) and an increase in elevation to 1120.23 feet, 0.73 feet above flood stage. Thus, we would conclude that we

```
EXECUTIVE CONTROL CARD    10 YR      OPERATION  INCREM,     MAIN TIME INCREMENT= 0.50              PASS=  1
EXECUTIVE CONTROL CARD                OPERATION  COMPUT,     FROM XSECTN/STRUCT  2/.0  TO XSECTN/STRUCT 28/.0
STARTING TIME= 0.0      RAIN DEPTH.= 1     RAIN DURATION= 1.00    RAIN TABLE NO.= 2      SOIL CONDITION= 2
ALTERNATE NO.= 1        STORM NO.= 1


SUBROUTINE RUNOFF    CROSS SECTION  2
            AREA= 1.13    INPUT RUNOFF CURVE= 70.0      TIME OF CONCENTRATION= 1.75

        PEAK TIMES          PEAK DISCHARGES         PEAK ELEVATIONS                 DRAINAGE AREA= 1.13
          13.11                218.128                  (RUNOFF)

 TIME              HYDROGRAPH, TZERO= 0.0                    DELTA T= 0.50
  0.0    DISCHG    0.0     0.0     0.0     0.0       0.0    0.0    0.0       0.0    0.0
  5.00   DISCHG    0.0     0.0                      40.15  191.46 137.92    102.10
 10.00   DISCHG   78.12    2.37                    154.38  216.14  28.31     26.53
 15.00   DISCHG   25.37   45.23                     36.01   33.24  18.15     16.21
 20.00   DISCHG   52.92   21.71                     20.56   19.70  18.73     10.01
 25.00   DISCHG   11.48    6.31    3.19    1.61      0.79    0.40   0.19      0.08   0.03

 TOTAL WATER, IN INCHES ON DRAINAGE AREA= 1.0390       CFS-HRS= 757.67     ACRE-FT= 62.61


SUBROUTINE SAVMOV    CROSS SECTION  2
             INPUT HYDROGRAPH= 6     OUTPUT HYDROGRAPH= 7


SUBROUTINE REACH     CROSS SECTION  3
             LENGTH= 1250.00    INPUT COEFFICIENT= 0.0     INPUT ROUTINGS= 0.0

    AVERAGE WATER VELOCITY= 0.130    AVERAGE ROUTING COEFF= 0.1000       NUMBER OF ROUTINGS= 0.54

        PEAK TIMES          PEAK DISCHARGES         PEAK ELEVATIONS
          14.81                88.996                 1116.66

 TOTAL WATER, IN INCHES ON DRAINAGE AREA= 1.0361       CFS-HRS= 755.60     ACRE-FT= 62.44


SUBROUTINE RUNOFF    CROSS SECTION  3
            AREA= 2.16    INPUT RUNOFF CURVE= 70.0      TIME OF CONCENTRATION= 2.72

        PEAK TIMES          PEAK DISCHARGES         PEAK ELEVATIONS
          13.82               305.553                  (RUNOFF)

 TOTAL WATER, IN INCHES ON DRAINAGE AREA= 1.0408       CFS-HRS= 1450.88    ACRE-FT= 119.90


SUBROUTINE ADDHYD    CROSS SECTION  5,6
             INPUT HYDROGRAPHS= 5,6    OUTPUT HYDROGRAPH= 7

        PEAK TIMES          PEAK DISCHARGES         PEAK ELEVATIONS
          13.95               385.356                 1119.16

 TIME              HYDROGRAPH, TZERO= 0.0                    DELTA T= 0.50                   DRAINAGE AREA= 3.29
  0.0    DISCHG    0.0     0.0     0.0     0.0       0.0    0.0    0.0       0.0    0.0
  5.00   DISCHG    0.0     0.0                      38.75  364.03 385.09    353.44
 10.00   DISCHG  300.87  258.97                    133.41 129.28 112.98    108.87
 15.00   DISCHG  245.46  265.46                    156.23 141.32  61.98     99.87
 20.00   DISCHG   45.79   35.73                     16.57  13.17  10.59     97.03
 25.00   DISCHG    5.79    1.96                     22.33   1.96   1.41      1.19
 30.00   DISCHG    1.01    0.72                      2.33   0.32   0.27      1.23
 35.00   DISCHG    0.19    0.14                      0.44   0.10   0.05      0.04
 40.00   DISCHG    0.04    0.03                      0.08   0.07   0.01      0.01
 45.00   DISCHG    0.01    0.0                       0.02   0.01

 TOTAL WATER, IN INCHES ON DRAINAGE AREA= 1.0392       CFS-HRS= 2206.48    ACRE-FT= 182.34
```

Figure 3-8. Model for runoff from watershed area (SCS hydrology model).

```
EXECUTIVE CONTROL CARD            OPERATION INCREM,     MAIN TIME INCREMENT= 0.50                          PASS= 1
EXECUTIVE CONTROL CARD   10 YR    OPERATION COMPUT,     FROM XSECTN/STRUCT 27, 0   TO XSECTN/STRUCT 28/ 0
STARTING TIME= 0.0      RAIN DEPTH= 3.55    RAIN DURATION= 1.00    RAIN TABLE NO.= 2    SOIL CONDITION= 2
ALTERNATE NO.= 1                  STORM NO.= 1

SUBROUTINE RUNOFF   CROSS SECTION 2                TIME OF CONCENTRATION= 1.75
           AREA= 1.13    INPUT RUNOFF CURVE= 70.0

   PEAK TIMES            PEAK DISCHARGES            PEAK ELEVATIONS
      13.11                 218.128                     (RUNOFF)

   HYDROGRAPH, TZERO= 0.0         DELTA T= 0.50         DRAINAGE AREA= 1.13

  TIME
  0.0    DISCHG    0.0    0.0    0.0    0.0    0.0    0.0
  5.00   DISCHG    0.0    0.0    0.0    0.0    0.0    0.0
 10.00   DISCHG   78.12  63.22  52.92  45.23  40.15  39.85
 15.00   DISCHG   25.37  24.15  21.71  21.10  37.92? 28.31
 20.00   DISCHG   11.48   6.31   ...
 25.00   DISCHG

   TOTAL WATER, IN INCHES ON DRAINAGE AREA= 1.0390    CFS-HRS= 757.67    ACRE-FT= 62.61

SUBROUTINE SAVMOV   CROSS SECTION 2
           INPUT HYDROGRAPH= 6    OUTPUT HYDROGRAPH= 7

SUBROUTINE REACH    CROSS SECTION
           LENGTH= 1250.00   INPUT COEFFICIENT= 0.0   INPUT ROUTINGS= 0.0   NUMBER OF ROUTINGS= 0.54
           AVERAGE WATER VELOCITY= 0.130   AVERAGE ROUTING COEFF= 0.1000

   PEAK TIMES            PEAK DISCHARGES            PEAK ELEVATIONS
      14.81                  88.996                     1116.56

   TOTAL WATER, IN INCHES ON DRAINAGE AREA= 1.0361    CFS-HRS= 755.60    ACRE-FT= 62.44

SUBROUTINE RUNOFF   CROSS SECTION 3                TIME OF CONCENTRATION= 2.72
           AREA= 2.16    INPUT RUNOFF CURVE= 90.0

   PEAK TIMES            PEAK DISCHARGES            PEAK ELEVATIONS
      13.59                 827.888                     (RUNOFF)

   TOTAL WATER, IN INCHES ON DRAINAGE AREA= 2.4996    CFS-HRS= 3484.44   ACRE-FT= 287.95

SUBROUTINE ADDHYD   CROSS SECTION 3
           INPUT HYDROGRAPHS= 5,6    OUTPUT HYDROGRAPH= 7

   PEAK TIMES            PEAK DISCHARGES            PEAK ELEVATIONS
      13.64                 899.437                     1122.10

   HYDROGRAPH, TZERO= 0.0         DELTA T= 0.50         DRAINAGE AREA= 3.29

  TIME
  0.0    DISCHG    0.0    0.0    0.0    0.0    0.0    0.0
  5.00   DISCHG    0.08   0.46   1.48   9.87   6.16   3.35
 10.00   DISCHG   47.16  37.99 102.48 758.13 475.82 283.07
 15.00   DISCHG  452.57 115.13 310.00 200.42 227.45 263.06
 20.00   DISCHG  125.17  44.37 108.01  92.20  96.84 101.90
 25.00   DISCHG   51.94   1.94  33.17  94.66  18.88  20.72
 30.00   DISCHG    1.01   0.86   0.72   0.61   0.44   0.52
 35.00   DISCHG    0.19   0.14   0.12   0.11   0.08   0.10
 40.00   DISCHG    0.04   0.03   0.03   0.02   0.02   0.02
 45.00   DISCHG    0.01   0.01   0.01
 50.00   DISCHG

   TOTAL WATER, IN INCHES ON DRAINAGE AREA= 1.9969    CFS-HRS= 4239.99   ACRE-FT= 350.39
```

Figure 3-9. Model for runoff from watershed (SCS hydrology model) after increase in cross-section 2 from 70.0 (Figure 3-8) to 90.0.

would have to disallow the land use change or force the developer to control the increase in runoff stemming from the change in order to avoid downstream flooding problems.

In many local ordinances, a different reference storm, such as the 100 year storm, might be used. We should also point out that the example run is not entirely correct since a major land use change would cause both the CN and the time of concentration to change. Again, this change must be calculated exogenously to the model.

It might be useful here to point out some of the problems associated with the implementation and interpretation of this model on the computer. First, the documentation is very cryptic and difficult to read. It assumes that the user has knowledge of the methods used from the other two background documents. Thus, it is recommended that the latter be read before an attempt to use the computer program. The documentation used for running the program itself—setting up the datasets, running the model, and so on— comes from the "give them an example and let them work through it" school of computer documentation. The text jumps from one item to the next, focusing on the mechanics of filling out the keypunching forms correctly rather than laying out the reasons and methods of constructing an entire run and interpreting the output. One must constantly flip back and forth from a terse reference such as "The 'Subroutine Operations' (Columns 4 through 12, Exhibit 16) are described by coded name and a number. The coded names, RUNOFF, RESVOR, etc. were previously described." The previous description appeared four pages earlier consisting of one or two lines while Exhibit 16, reproduced below as Figure 3-10, is about 15 pages further ahead. A look at this exhibit prompts the following questions:

1. Must REACH always precede the second runoff computation or is order not important here?
2. What are the input and output numbers shown on several of the statements (an attempt is made to explain this about two pages after the reference to Exhibit 16).
3. Where does one get the data?
4. Where and how are these instructions reflected on the output? The output is given later in the manual without any explanation, leaving the user to correlate the output with the input instructions.

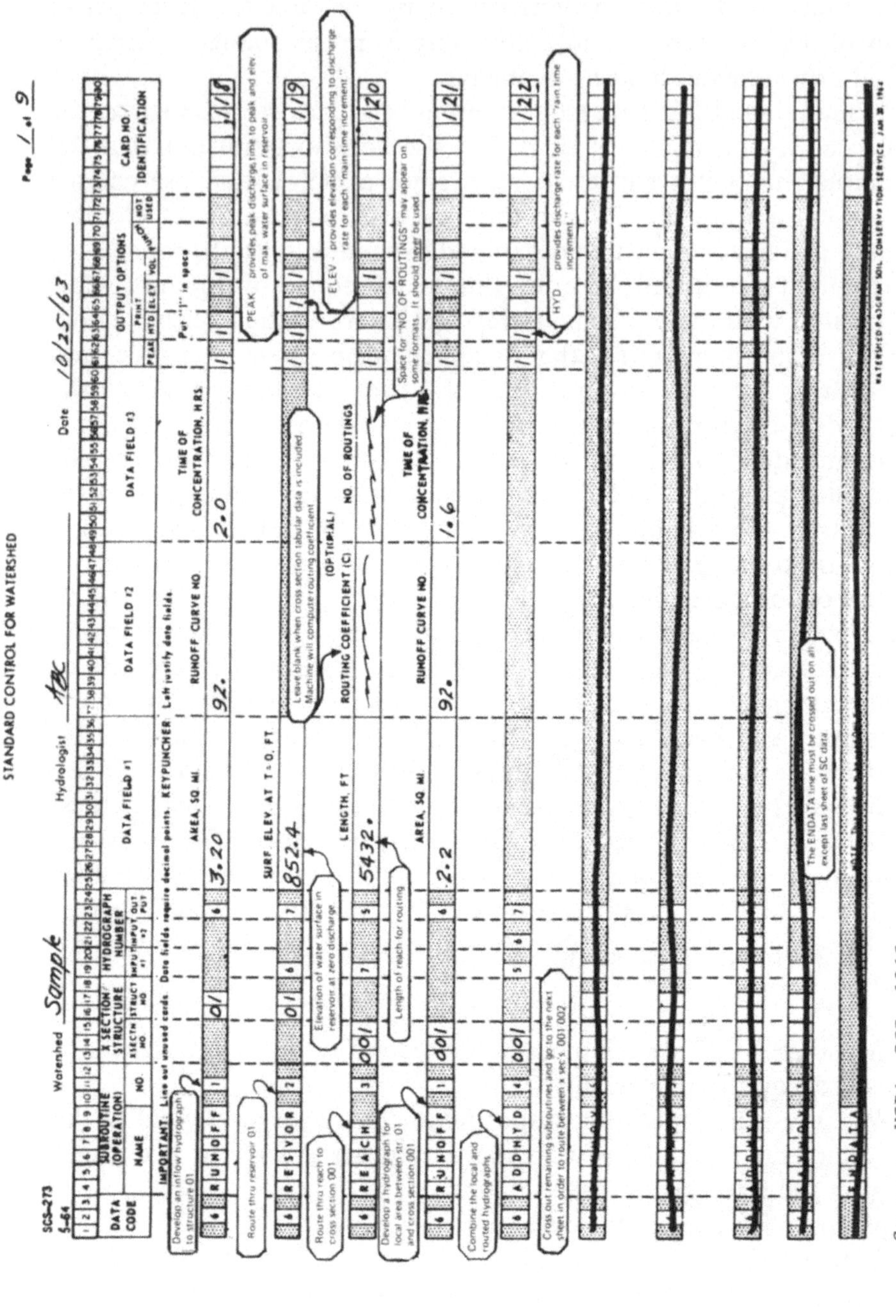

Figure 3-10. Standard control for watershed. (Source: USDA, SCS, 1965.)

Source: USDA, SCS, 1965.

A rewriting of the model documentation is certainly beyond the scope of this book. Some problems might be avoided by keeping the following in mind:

1. Most of the input data sources and exogenous calculations are discussed in Technical Release 55 and the *National Engineering Handbook.*
2. Flood stages are not explicitly set in the model but are only implied via the bankfull stage as the final data item in the description of reservoir (structure) and cross-section elevation/flow relationships. Flooding is not explicitly modeled but must be implicitly concluded with the same information as in our example.
3. Following the input of tables representing rainfall, structures, and cross-sections, the model utilizes two sets of control instructions. The first creates the rainfall and runoff and routes it step by step through each cross-section and structure until it reaches the other end of the basin. The second set then gives the go-ahead to compute through the first sequence and gives the computer information on the storm conditions that are to be simulated. A third set of instructions is optional, causing alterations in the first two to be run.

We might point out that once the basic principles associated with the model become familiar to the user, a process that might take a day, the model is relatively straightforward.

With the second example, the STORM model, the model input is relatively easy to decipher. However, the interpretation of the output can be problematic. Upon our first use of the model, it took two days to find the key to the abbreviations table for the output. Even with this in hand, the interpretation was not easy. The major water quantity-quality analysis table is shown as Figure 3-11. The key is shown as Figure 3-12.

The major difference to bear in mind while reviewing this model is that it simulates only the impact of dry weather and stormwater flows for one storm and *for one point in a watershed.* The point can be a critical stream cross-section, pipe section, or reservoir. It can have mixture of land uses and soil types on its watershed area. However, only a small watershed can be represented in this way because no

PAGE 1   TEST DATA SET 2

BOISE, IDAHO

TREATMENT RATE = .0010 IN/HR.   6 CFS,   .391 MGD
STORAGE CAPACITY = .0200 INCHES,   1.0 AC = FT,   .326 MG

QUANTITY ANALYSIS

WEATHER BUREAU, BOISE AIRPORT
BOISE NO. 2

EVENT	YEAR	MO	DY	HR	HRS NO STORAG	RAINFALL DRIN	RAINFALL HRS	RAINFALL INCH	RUND INCH	OUTF INCH	HRS TO EMPTY	STORAGE DURTN	STORAGE MAX	NO	8T	OVERFLOW OUR	OVERFLOW WASTE	OVERFLOW INTL	TREATMENT HRS	TREATMENT INCH	AGE1	AGE2	AGE3	AGE4	AGE5
1	72	1	9	11	273	42	14	.53	.11	.11	0	42	.02	1	5	6	.07	.04	55	.04	0	0	0	0	0
2	72	1	11	8	3	27	11	.28	.05	.05	9	36	.02	2	3	4	.01	.01	38	.03	0	0	0	0	0
3	72	1	18	5	129	11	10	.53	.15	.15	24	35	.02	3		11	.11	.04	38	.11	0	0	0	0	0
4	72	1	20	5	13	75	37	1.53	.62	.62	22	97	.02	4	8	35	.52	0	99	.09	0	0	0	0	0
5	72	2	15	13	535	28	2	.09	.02	.02	16	18	.01		5	NO OVERFLOW			38	.02	0	0	0	0	0
6	72	2	28	5	288	28	7	.50	.07	.07	14	46	.02	5	5	5	.02	.01	86	.04	0	0	0	0	0
7	72	3	2	3	24	15	12	.88	.45	.45	24	39	.02	6	5	13	.41	0	41	.03	0	0	0	0	0
8	72	3	18	12	354	6	6	.58	.24	.24	24	30	.02	7	2	9	.21	.08	51	.03	0	0	0	0	0
9	72	4	11	21	555	1	1	.16	.01	.01	9	10	0		4	NO OVERFLOW			36	.01	0	0	0	0	0
10	72	4	12	18	11	4	3	.05	.01	.01	6	10	0		4	NO OVERFLOW			11	.01	0	0	0	0	0
11	72	4	16	12	81	1	1	.05	.01	.01	9	10	0		4	NO OVERFLOW			13	.01	0	0	0	0	0
12	72	5	21	21	839	1	1	.03	.00	.00	4	5	0		4	NO OVERFLOW			46	.01	0	0	0	0	0
13	72	6	6	14	373	2	2	.36	.06	.06	24	26	.02	8	2	4	.03	.02	28	.02	0	0	0	0	0
14	72	6	7	19	3	17	5	.38	.07	.07	11	28	.02	9	4	4	.03	.03	28	.02	0	0	0	0	0
15	72	6	9	3	4	18	3	.03	.01	.01	3	6	0		3	NO OVERFLOW			8	0	0	0	0	0	0
16	72	6	9	24	15	18	2	.22	.03	.03	8	26	.02	10	4	2	0	0	27	.02	0	0	0	0	0
17	72	7	21	18	976	11	7	.27	.03	.03	24	26	.02	11	11	2	.05	.04	28	.02	0	0	0	0	0
18	72	9	5	6	1066	10	7	.51	.08	.08	24	35	.02	12	4	4	.02	.02	42	.03	0	0	0	0	0
19	72	9	11	7	110	10	6	.27	.05	.05	17	27	.02	13	4	3	.03	.02	29	.02	0	0	0	0	0
20	72	9	27	3	357	5	5	.27	.06	.06	24	29	.02	14	5	4	.03	.02	49	.02	0	0	0	0	0
21	72	10	9	20	276	7	5	.20	.04	.04	21	28	.02	15	5	3	.01	0	29	.02	0	0	0	0	0
22	72	10	19	16	208	4	3	.11	.02	.02	12	16	0		6	NO OVERFLOW			28	.01	0	0	0	0	0
23	72	11	4	5	357	22	8	.42	.06	.06	18	40	.02	16	5	5	.02	.02	45	.04	0	0	0	0	0
24	72	11	8	4	55	6	5	.31	.03	.03	23	29	.02	17	7	2	0	0	29	.02	0	0	0	0	0
25	72	12	18	10	938	48	26	1.63	.65	.65	9	57	.02	18	4	25	.59	.05	115	.06	0	0	0	0	0
26	72	12	22	16	47	1	1	.05	.01	.01	6	7	0		3	NO OVERFLOW			17	0	0	0	0	0	0
27	72	12	23	23	24	7	7	.28	.07	.07	24	31	.02	19	4	7	.04	.02	32	.03	0	0	0	0	0
AVE OF 27 EVENTS					293.1 **	13.9	7.1	.39	.11	.11	15.3	29.2	.02		4.0*				40.2	.03	0	0	0	0	0
AVE OF 19 OVRFLW EVENTS						18.9	9.4	.53	.15	.15	18.3	37.2	.01*		4.6	7.3	.12	.03	46.8	.04	0	0	0	0	0

NON-OVERFLOW EVENTS ONLY,
EXCLUDING 8 DRY PERIODS

Source: U.S. Army Corps of Engineers, 1977.

Figure 3-11. Water quantity/quality analysis table (STORM). (Source: U.S. Army Corps of Engineers, 1977.)

DEFINITIONS OF QUANTITY COLUMN HEADINGS

1	EVENT	= SEQUENCING NUMBER.
2	DATE	= DATE THIS EVENT BEGAN.
3	HR	= NUMBER OF HOURS PAST MIDNIGHT THIS EVENT BEGAN.
4	HRS NO STORAG	= NUMBER OF HOURS SINCE END OF LAST EVENT, EXCLUDING SUMMER (MORE THAN, 1440 HOURS).
5	DRTN	= DURATION OF STORM FROM FIRST HOUR OF RAIN, TO LAST HOUR OF RAIN.
6	HRS	= NUMBER OF HOURS IN WHICH RAINFALL OCCURRED DURING EVENT.
7	INCH	= AMOUNT OF RAINFALL DURING THE EVENT IN INCHES.
7A	RUNO INCH	= SURFACE RUNOFF DURING EVENT IN INCHES.
7B	OUTF INCH	= TOTAL OUTFLOW (SURFACE RUNOFF + DRY WEATHER FLOW).
8	HRSTO EMPTY	= NUMBER OF HOURS FROM LAST RAINFALL TO END OF EVENT.
9	DURIN	= TOTAL NUMBER OF HOURS STORAGE WAS UTILIZED. IE, LENGTH OF THE EVENT.
10	MAX	= MAXIMUM AMOUNT OF STORAGE UTILIZED, IN INCHES.
11	NO	= OVERFLOW EVENT SEQUENCING NUMBER.
12	ST	= NUMBER OF HOURS ELAPSED BEFORE OVERFLOW STARTED. OR, IF NO OVERFLOW, HOUR OF MAXIMUM STORAGE.
13	DUR	= NUMBER OF HOURS IN WHICH OVERFLOW OCCURRED.
14	WASTE INCH	= QUANTITY OF WATER RELEASED UNTREATED, IN INCHES.
15	INTL	= QUANTITY OF WATER RELEASED UNTREATED DURING THE FIRST 3 HOURS OF OVERFLOW.
16	HRS	= NUMBER OF HOURS WATER WAS TREATED DURING THE PRESENT EVENT AND SINCE THE PREVIOUS EVENT.
17	INCH	= QUANTITY OF WATER TREATED DURING THE EVENT AND SINCE THE PREVIOUS EVENT.
18	AGE1	= AVERAGE AGE (HOURS) OF TREATED RUNOFF.
19	AGE2	= MAXIMUM AGE (HOURS) OF STORAGE ON FIRST IN, FIRST OUT BASIS.
20	AGE3	= MAXIMUM AGE (HOURS) OF STORAGE ON FIRST IN, LAST OUT BASIS.
21	AGE4	= QUANTITY WEIGHTED AVERAGE AGE (HRS) OF STORAGE ON FIRST IN, FIRST OUT BASIS.
22	AGE5	= QUANTITY WEIGHTED AVERAGE AGE (HRS) OF STORAGE ON FIRST IN, LAST OUT BASIS.

Figure 3-12. Key to water quantity/quality analysis table of Figure 3-11 (STORM).

routing mechanism is included in the STORM model. The model output shows in some detail the quantity and quality impacts of land use changes, for the representative storm used, but only calculates either the peak flow or the unit hydrograph flow through the point in question. It explicitly represents the overflows associated with given rainfalls and in this respect is easier to use than the SCS model.

Going back to the example output, Figure 3-11, we can see how these calculations are implemented in the program. At the top left, we see that the storage capacity for this stream segment or basin is 0.200 inches. This translates to 1.0 acre-feet or 0.6 cfs. This is the storage capacity that is used to determine whether the storm will produce an overflow and by how much. The storm data are given in the first several columns of the output. The sequence number on the far left is followed by the year, month, day, and hour. Then, following the information on previous storms come data on the rainfall and runoff, and storage. The most critical column relative to stormwater quantity is the next set of numbers, labelled "Overflow." This column assigns a sequence number to each overflow event and shows the extent and time of occurrence of the overflow. For example, event number 1 started after five hours, lasted for six hours and released .07 inches of water untreated. The worst overflow for this storm sequence occurs as overflow 18 and results in a 25 hour overflow of .59 inches. This converts to 31.47 acre-feet or 10.25 million gallons of flood waters.

By conducting this analysis for several different storm probability events, one could obtain the flood analysis for a series of storms. By making changes in the percentages of developed land and(or) time of concentration, one could simulate the impacts of land use changes on storm water runoff and therefore the total amount of water that must be stored to meet any local drainage regulations.

REFERENCES

Espey, W. H. and D. E. Winslow (1975). *Quantity Aspects of Urban Stormwater Runoff,* Short Course Proceedings, Applications of Stormwater Management Models, National Environmental Research Center, Office of Research and Development. Cincinnati, OH: U.S. Environmental Protection Agency.

Henderson, F. M. (1963). "Some Properties of the Unit Hydrograph," *Journal of Geophysical Research,* Vol. 68, p. 4785–4793.

Horton, R. E. (1945). "Erosional Development of Streams and Their Drainage Basins:

Hydrophysical Approach to Quantitative Morphology," *Bulletin of the Geological Society of America,* Vol. 56, p. 275–370.

Linsley, Ray K., Jr., Max A. Kohler, and Joseph L. H. Paulhus (1975). *Hydrology for Engineers.* New York: McGraw-Hill.

Maloney, Frank E., Richard G. Harmann, and Brian D. E. Carter (1980). "Stormwater Runoff Control: A Model Ordinance for Meeting Local Water Quality Management Needs," *Natural Resources Journal,* Vol. 20, No. 4, p. 713–764.

Miller, William A. and Jean A. Cunge (1975). "Simplified Equations for Unsteady Flow," in *Unsteady Flow in Open Channels,* Vol. I, ed. by K. Mahmood and V. Yevjevich. Fort Collins, CO: Water Resources Publications. P. 183–258.

Overton, Donald E. and Michael E. Meadows (1976). *Stormwater Modeling.* New York: Academic Press.

Sherman, L. K. (1932). "Streamflow from Rainfall by the Unit-Graph Method," *Engineering News Record,* Vol. 108 (April 7).

Urban Land Institute (1975). *Residential Stormwater Management.* Washington, DC: Urban Land Institute.

U.S. Army Corps of Engineers, Hydrologic Engineering Center (1977). *Storage, Treatment, Overflow, Runoff Model "STORM" User's Manual.* 723-58-17520. Davis, CA: HEC, U.S. Army Corps of Engineers.

USDA: Agricultural Handbook (1965). *Rainfall-Erosion Losses from Cropland East of the Rocky Mountains,* Agricultural Handbook 202. Washington, DC: U.S. Government Printing Office.

USDA, SCS (1965). *Computer Program for Project Formulation Hydrology.* Technical Release 20. Washington, DC: USDA.

USDA, SCS (1972). "Hydrology," Sec. 4 in *SCS National Engineering Handbook.* Washington, DC: U.S. Government Printing Office.

USDA, SCS (1975). *Urban Hydrology for Small Watersheds.* Technical Release 55. Washington, DC: USDA.

Ward, R. C. (1967). *Principles of Hydrology.* New York: McGraw-Hill Book Co.

Yevjevich, Vujica (1975). Introduction to *Unsteady Flow in Open Channels,* Vol. I, ed. by K. Mahmood and V. Yevjevich. Fort Collins, CO: Water Resources Publications. P. 1–28.

4
Air Pollution Models

INTRODUCTION

In parallel with the Clean Water Act, the Clean Air Act and its amendments have established standards relating to the reduction of air pollution emissions from existing and new sources. Regulations exist for many pollutants but much of the research and impact simulation has focused on six major pollutants: sulfur oxides, particularly SO_2; nitrogen oxides, including NO_2, NO_3 and others; total suspended particulates or TSP; hydrocarbons (HC); carbon monoxide (CO); and photochemical oxidants, particularly ozone. The Clean Air Regulations are rather complex, involving a number of different standards as well as multiple approvals for new sources of air pollution. It might be useful to give a brief overview of these regulations since the models used to describe air pollution impacts are so heavily influenced by them. Following this review, the basic theories behind the air pollution models will be discussed and followed by specific model reviews.

The air pollution regulations presently (in 1984) include standards for both ambient and effluent conditions. Ambient standards are set for the general environment based on the impacts that excessive pollutants would have on human and ecological health. The ambient standards for the six major pollutants are shown as Table 4-1. These are called the *primary standards,* a set of less stringent standards now in force that were originally to be followed by a more stringent set of secondary standards. At the time of this writing, though, even the less stringent standards are under review.

The ambient environment is protected through the promulgation of effluent standards. These regulate the rate of emissions from old and new sources of air pollution. These sources could be *point sources,* coming from an individual location such as the smoke stack from an industrial boiler; *line sources,* coming from major highways; or *area sources,* coming from multiple, small points such as houses,

Table 4-1. National Ambient Air Quality Standards

(Established Pursuant to the Clean Air Act of 1970)

Pollutant	Period of Measurement[a]	Primary Standard		Secondary Standard	
		$\mu g/m^{3[b]}$	ppm[b]	$\mu g/m^{3[b]}$	ppm[b]
1. Carbon Monoxide (CO)	8 hours	10,000	9	Same	Same
	1 hour	40,000	35	Same	Same
2. Hydrocarbons (HC) (non-methane)	3 hours	160	0.24	Same	Same
3. Nitrogen Oxides (NO_2)	Year	100	0.05	Same	Same
4. Photochemical Oxidants (O_x)	1 hour	160	0.08	Same	Same
5. Sulfur Oxides (SO_x)	Year	80	0.03	None	None
	24 hours	365	0.14	None	None
	3 hours	None	None	1,300	0.5
6. Total Suspended Particulates (TSP)	Year	75	—	60	—
	24 hours	260	—	150	—

[a]Concentrations are averaged over each period of measurement. The annual TSP concentration is a geometric mean of 24-hour samples; all other concentrations are arithmetic mean values. Standards for periods of 24 hours or less may not be exceeded more than once per year.

[b]Units of measurement are micrograms per cubic meter ($\mu g/m^3$) and parts per million (ppm).

Source: USEPA, 1974

residential streets, and so forth, adding up to significant sources. Only point and line sources are directly regulated by the Environmental Protection Agency.

The regulations require the emissions, whether from individual smoke stacks or motor vehicles, to be below certain rates of pollutant production. The emission rates are set such that, if they are met, the ambient air is supposed to achieve its required standards. Different effluent standards are set for existing firms and new firms. Existing firms must install pollution control devices that are somewhat less expensive to use—called the Best Practicable Technology (BPT); while new firms must use the Best Available Current Technology (BACT).

In practice, air quality is reviewed in geographic regions, generally one or more counties, called Air Quality Control Regions (AQCR's). In areas where the ambient standards are not being met, a plan for controlling pollution must be promulgated. Since this plan is generally made on a state by state basis, it is called the State Implementation Plan (SIP). For areas with air quality that is already good, the standards are stricter to prevent the deterioration of that quality. These standards are called the Prevention of Significant Deterioration Standards (PSD).

The upshot of this alphabet soup of regulations comes when one wishes to locate a new plant, build a new highway, and so on. An analysis is required showing that the air pollution impacts will not violate the regulations. If the increment of new pollution will exceed either the PSD or ambient standards, then the project must be cancelled or a pollution offset must be obtained which is greater than the new pollution to be introduced. An offset is achieved by getting some existing polluter either to change or shut down its operation. There are in fact examples in which companies have sold their "pollution rights" to new industries and thus agreed to shut down or dramatically alter their operations. There are also cases in which the new industry has acquired an older plant and modernized it, thus reducing air pollution and producing the requisite offset.

Given the fact of these regulations, the importance of the use of air pollution models should be apparent. Without an analysis of its air pollution impacts, any new company with a significant emission level can be prevented from operating. Because the regulations are interpreted by the U.S. Environmental Protection Agency, most of the models have been developed for and used by that agency. As is the case with water pollution models, there are many specialized types of models aimed at analyzing various aspects of the pollution problem from different points of view. In air pollution, these include models aimed at the evaluation of the standards themselves, models used to establish a monitoring network, models for the prediction of pollution concentration, models focusing on changes in levels with time, and models which optimize air pollution distribution. The set which is used most frequently and has received the most attention in terms of computer software development is that associated with the prediction and forecasting of pollutant concentrations and the analysis of pollution trends. Thus, these models will be the focus of our further discussion. For a review of all the types of air pollution models, the reader is referred to an article by Singpurwalla (1975).

THEORETICAL CONSIDERATIONS

If we harken back to the confused raindrop analogy of Chapter 3 and add to the confusion a three dimensional framework that varies instantaneously with time, we can conceptualize the major problems associated with the modelling of air pollution. In the ambient environment, air movements can occur in any direction. The direction and velocity of movement can shift over very short time periods of a few seconds to a few minutes. Then adding to this complexity the variation in the emission rates for different pollutants over time, the various locations of emission sources and recipients, and the chemical processes associated with pollutant degradation, or conversion to other forms either directly or through synergisms with other pollutants, gives one an inkling of the difficulty of modelling air pollution.

In order to accommodate all of these and many other possible variations associated with the atmospheric environment, a number of simplifying assumptions must be made. Several concepts are critical to an understanding of these assumptions. First, one must understand a profound principle of atmospheric physics: hot air rises. This phenomenon influences air movement in both the vertical and horizontal directions.

Under normal circumstances, the temperature of air declines with altitude at a constant rate, called the *adiabatic lapse rate,* of 5.5° F per 1000 feet of altitude. This is represented diagramatically by Figure 4-1. Under such circumstances, a warm parcel of air, as from an industrial smokestack, would rise until it cools off enough to reach an equilibrium with the surrounding air. Under some circumstances, the local lapse rate is higher than normal. Because of this, air rising through the parcel will move much more quickly than under "normal" circumstances and to a higher altitude. Such an air mass is thus said to be unstable. At the other extreme is an air mass in which the temperature change is the reverse of normal; this is the air gets warmer with altitude. Here, rising air parcels will tend to settle back toward the ground. This condition is said to be an *inversion* because of this inverse temperature condition; and because air movement is impeded, the air mass is said to be *stable*. It is under these circumstances that air pollution episodes occur, because of the restricted mixing of emissions in the atmosphere.

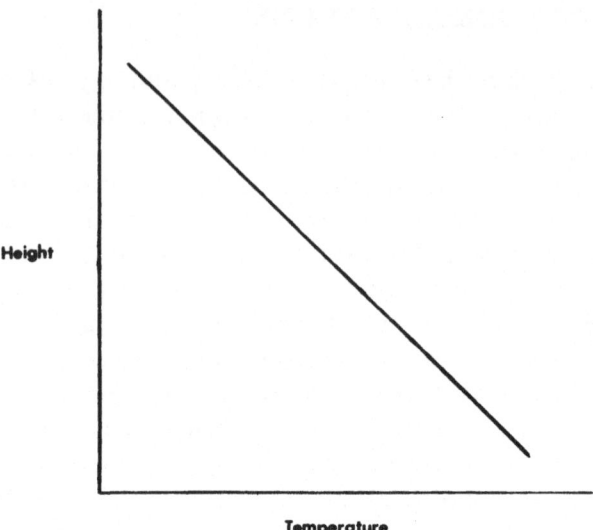

Figure 4-1. Normal temperature profile.

Air movement in the horizontal direction (the technical term is "wind") is caused by the differential heating and cooling of the earth's surface. Wind direction and speed is also influenced by the coriolis effect of the earth's rotation and by the friction of the surface. Because of the frictional forces, wind speed generally increases with altitude as a power function:

$$\frac{\mu}{\mu_1} = \left(\frac{z}{z_1}\right)^p$$

where

μ, μ_1 = wind speed
z = height
z_1 = reference height
p = exponent between 0 and 1

With an adiabatic lapse rate, level terrain, and low surface cover, p is $\frac{1}{7}$. Because of the friction of ground cover, air turbulence is greatest near the ground. Topographic barriers create eddies that change the direction and speed of the air; an example is given in Figure 4-2.

Figure 4-2. Topographic barriers and wind.

Movement of the air obviously affects the mixing of air pollutants with "clean" air. Measurement of the resulting pollution levels is by no means itself a simple concept. When one samples water for pollution levels, a finite amount of water is collected and tested for pollutant concentrations. Because the system is a channelized flow system, assumptions can be made concerning the sample as representative of the overall quality of the water.

With air quality, the potential constant movement in all directions has a greater impact on the nature of the measurements being made at any instant. Thus, the concept of averaging time becomes more critical. This can be illustrated by considering an air pollution monitor which takes a sample by pumping air through the testing chemicals for a period of five minutes. The reading thus obtained represents an average of the conditions during that five minute period. If the averaging time is doubled, the new air pollution levels associated with the ten minute interval can be computed as a function of the five minute readings by calculating the mean:

$$Y_1 = \frac{X_{t1} + X_{t2}}{2}, \quad Y_2 = \frac{X_{t3} + X_{t4}}{2}, \ldots$$

where

X_{ti} = the five minute averaging time concentrations;
Y = the new ten minute averaging time concentrations.

The result of the averaging is to reduce the reported maximum and minimum values by averaging them with other values. As the averaging time gets longer, the extremes of measurement or range are also reduced. Thus, averaging time greatly impacts the values of air pollution concentration. It is for this reason that the air quality standards shown in Table 4-1 include explicit averaging times.

PRINCIPAL MODELS

Box Model

Given these simple concepts, we can begin to discuss some of the approaches to modelling air pollution that have been given the greatest attention. The most simplistic representation of air pollution diffusion that has gained some degree of use is called the *box model.* It cannot be used for the types of analyses needed to meet the requirements of the air pollution regulations because it is highly inaccurate, but it is often used to get a first approximation of relative air pollution levels for different proposals. Its advantage is that the calculations can be made on a calculator. The model makes the following assumptions:

1. The atmosphere around the area being evaluated can be represented by a box of known dimensions.
2. The emission rate S for the pollutant(s) of interest is constant over time.
3. There is instantaneous and perfect mixing of the air within the box giving a uniform air pollution concentration.
4. The pollutants do not react chemically.
5. Only clean air enters the box while polluted air leaves the box at the same rate (Q).

The equations for this model then become (Singpurwalla, 1975):

$$Q = hLu$$

where

    Q = the rate of airflow into and out of the box;
    u = mean wind speed;

h = inversion height i.e. the height of the box;
L = length of the box.

If C(t) is the concentration of a pollutant at time t then the mass balance for the pollutant in the box is:

$$\frac{dc}{dt} = \frac{S}{V} - \frac{C}{T}$$

where $T = V/Q$ is the average residence time of a unit mass of pollutant in the box. Then, if C(0) is the pollutant concentration at time = 0, C(t) has the solution

$$C(t) = \frac{S}{Q}(1 - e^{-t/T}) + C_0 e^{-t/T}$$

Over time, given the assumptions of steady state conditions, a steady state concentration of pollutant will be reached.

Obviously, this model leaves much to be desired. None of the variations in meteorological conditions that are known to exist are taken into account. In addition, only simple point sources of air pollution can be considered and multiple sources are difficult to handle. Thus, the model is really only used as a general indicator of the air pollution impacts of a single source in rather simple circumstances.

Gaussian Plume Model

The model which is most frequently used as the basis for more complex air pollution calculations is the *Gaussian plume model.* This model assumes that over short periods of time—say one hour or less—steady state conditions exist with regard to air pollution emissions and meteorological changes. Conceptually, the model can be understood by looking at Figure 4-3. Here, one can see that the air pollution is represented as an idealized plume coming from the top of the stack. Because the plume is being produced by a fuel burning

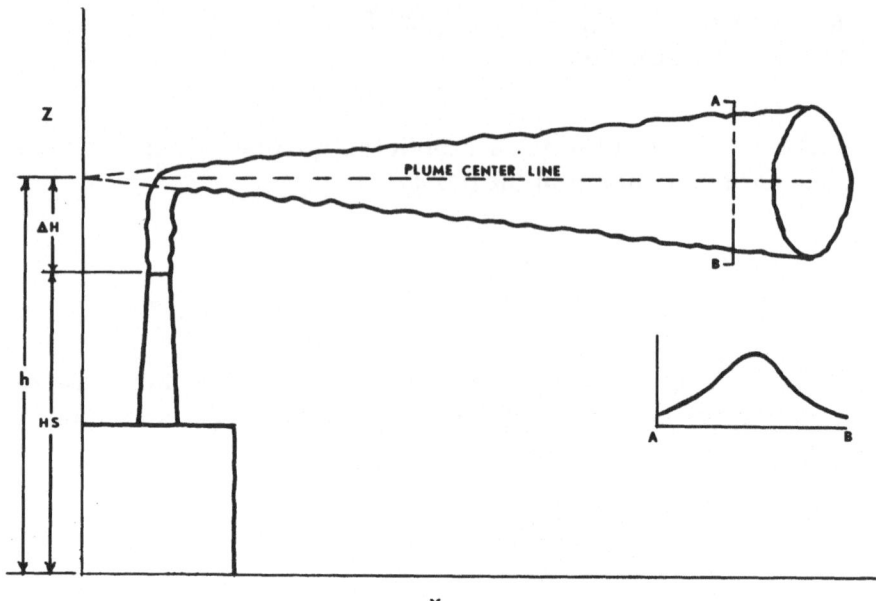

Figure 4-3. Idealized representation of air pollution: the Gaussian plume.

process, the plume will be thrust upward some distance Δh between the top of the stack, of height H, and the effective stack height h. The actual value of this movement will depend on stack gas exit velocity and on temperature as it relates to the temperature of the surrounding air.

Once the plume has reached the effective stack height, dispersion will begin in three dimensions. Dispersion in the downwind direction x will be proportional to the mean wind speed and in that direction. Dispersion in the cross-wind direction y and in the vertical direction z will be governed by the Gaussian plume equations. In effect, this model assumes that dispersion in these two directions will take the form of a normal or Gaussian curve, with the maximum concentration in the center of the plume, called the *plume center line,* and decreasing concentrations toward the edges of the plume. In actual practice, the solution of the model involves a semi-empirical representation of a set of partial differential equations converted into difference equations for solution on a digital computer. The key factors in this equation are the coefficients of dispersion in the y and z dimensions. This will be discussed further below. Now, it remains to give the basic Gaussian plume equation:

$$c(x, y, z) = \frac{Q}{2\pi \overline{u}\sigma_y\sigma_z} \exp\left(-\frac{y^2}{2\sigma_y^2}\right) \left[\exp\left(-\frac{(z - h)^2}{2\sigma_z^2}\right)\right.$$

$$\left. + \exp\left(-\frac{(z + h)^2}{2\sigma_z^2}\right)\right]$$

where

$C(x, y, z; h) =$ pollutant concentration at point x, y, z for an effective stack height h, $\mu g/m^3$;

$Q =$ emission rate, gms/sec;

$u =$ mean wind speed, meters/sec;

$\sigma_y, \sigma_z =$ standard deviation of the plume concentration distribution along Y and Z axis respectively. (Note: σ_y and σ_z are functions of downwind distance and atmospheric stability.)

The dispersion coefficients vary with downwind distance and stability class. The more unstable the air, represented as class A, the higher the dispersion coefficient. Figure 4-4 shows the relationship between stability class, distance from the source, and horizontal and vertical dispersion imbedded in most Gaussian plume models.

In operation, the model is calculated on a cell by cell basis using a difference representation of the plume model. At the end of each time-distance increment, the concentration of the pollutant in the grid cell can be calculated. This pollutant is then passed to the next cell and dispersed further and so on until some user-defined distance downwind. At selected points downwind, the user specifies a set of receptors at which the final, ground level pollution levels are reported. In this way, one calculates the steady state pollution levels caused by a specific set of meteorological conditions by a single, constant pollution source for pollutants that are not reactive in the environment.

In order to make the analysis more realistic, other conditions must be taken into account. In particular, one must account for the variation in meteorological conditions over time. This is represented

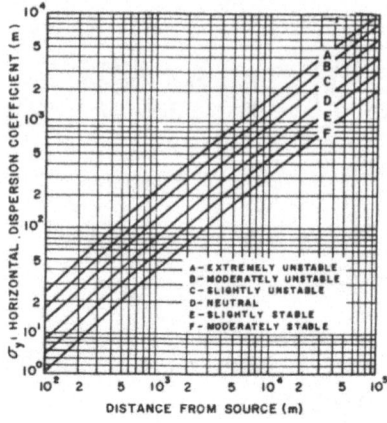 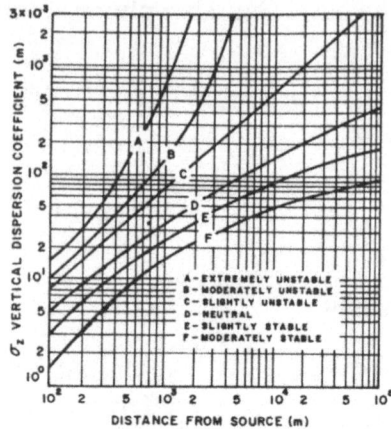

Figure 4-4. Relationships embedded in Gaussian plume models. Lateral standard deviation (left) and vertical standard deviation (right) of diffusing plumes as functions of downwind distance and stability category. (Source: USAF, Air Weather Service, 1971).

using a probability distribution showing the variations in wind speed and direction for each stability category. The composite diagram showing these data in graphic form is called the *stability wind rose.* An example is given as Figure 4-5. The computerized versions of the Gaussian plume models use selected wind directions, speed classes, and the stability categories entered as a probability matrix to calculate the steady state pollutant concentration under each condition. The conditions are then averaged using the probabilities as weights to derive the annual average pollutant concentrations at each of the receptors selected. Other calculations, to be discussed below, may be used to calculate the concentrations for shorter averaging times.

Multiple sources of air pollution must be accommodated by calculating the separate Gaussian plumes and adding them together as the same grid cells are encountered. Further model alterations must be made to accommodate intermittent sources, line sources, and area sources of air pollution.

Although many of the initial assumptions of the Gaussian plume model can be relaxed through additional calculations, it still generalizes the dispersion of air pollutants to a great degree. There are many meteorological conditions under which air pollution plumes are not Gaussian (see Stern, 1976). In addition, the calculations used to estimate shorter averaging times, to accommodate line and area sources, and to provide for other than ideal conditions all introduce calcula-

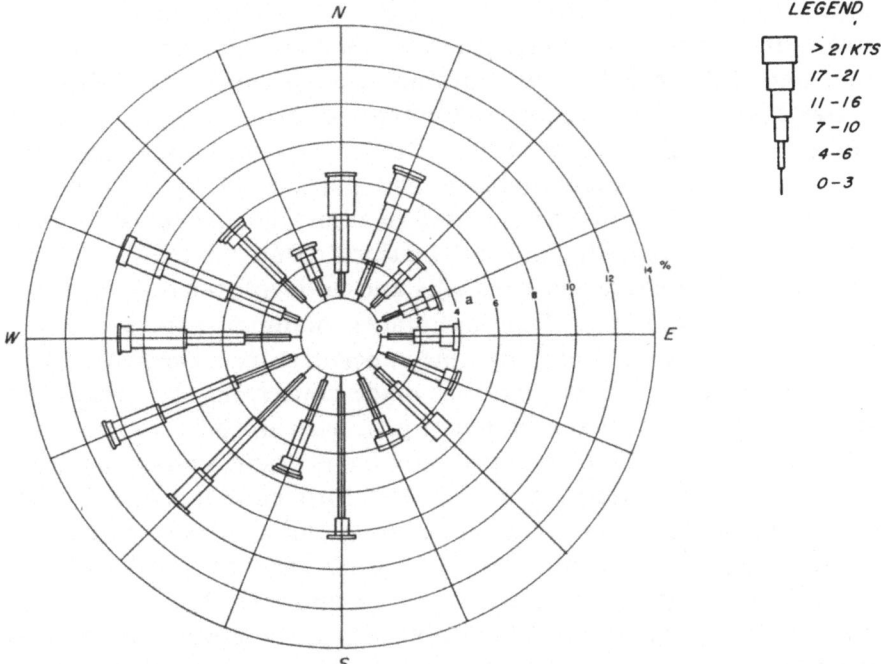

Figure 4-5. The stability wind rose for Newark (N.J.) International Airport. Concentric percentage circles indicate frequency of occurrence (e.g., easterly winds occur at 4 percent of the time during the year. (Source: USEPA, 1974.)

tion errors of their own. These conditions are further complicated by the very sparse air pollution monitoring network which makes comparison of modelled and actual levels difficult. Nevertheless, attempts to validate the model have shown it to be correct only within a factor of 2 to 10 times. Yet, it remains the accepted method for calculating the air pollution impacts of all types of new developments because it is theoretically sound in approach and because it represents one of the few available tools of practical value.

COMPUTER MODEL DESCRIPTIONS

The Climatological Dispersion Model

One of the most frequently used models employing the Gaussian plume model to simulate the impacts of point and area sources of air pollution is the Climatological Dispersion Model (CDM) and its update CDMQC (USEPA, 1973; USEPA, 1977). This model is relatively straightforward to use and interpret and is extremely well documented. Table 4-2 shows the card input requirements for the model.

Basically the model requires nine types of data for input. The first and second cards have identifying information about the run being made, input and output datasets and parameters for an optional regression equation that can be used to adjust the model results to an empirically derived adjustment. The third card completes the information on the calibration constants and adds information on background concentrations of the pollutants, pollutant names, and options for alternative averaging time concentrations. The fourth card involves the data and assumptions made concerning the geographic scale and location of the mapping grid used in the computer run, several conversion factors, calculation factors, and ambient environment data assumptions. Similarly the fifth card involves further definition of the assumptions concerning emission rates and decay half-lives for reactive pollutants. In each case, the card must be included but many of the parameters are optional, that is if they are not defined by the user a default option is automatically inserted by the model.

The next set of data cards show the joint probability distribution for the stability wind rose for the location being modeled. In most cases, these data can probably be obtained from the State or USEPA who have run a model such as CDM for most areas in the United States. In cases in which for some reason these data are not available, one must obtain the three hour meteorological data for the area in question and calculate the frequency distribution using a special computer subroutine. Such data are available for all major airports in the U.S. from the National Oceanic and Atmospheric Agency in Asheville, North Carolina. For areas outside of the U.S., one would have to calibrate a similar frequency distribution based on airport meteorological records for the area in question. The CDM model was in fact tested in a foreign context using Ankara, Turkey, for the test.

Following this set of cards, one must input the point source emissions inventory. On each card, one indicates the location of the source on the grid system, estimates of stack height, emission rates, and so on. Two pollutants can be simulated in the same run. These are the most difficult data to assemble if they have not already been

Table 4-2. Card Input to CDMQC[a,b]

Card No.	Column	Format	Contents
1	1–80	20A4	*ITIT (1)–ITIT (20) (Title of run to be printed at top of every page of output)
2	1–8	2A4	AROS (1)–AROS (2) (Identification for punched output of the computed area source concentrations of the two pollutants)
	9–16	2A4	PROS (1)–PROS (2) (Identification for punched output of the computed point source concentrations of the two pollutants)
	17–21	I5	IRUN (Computer run identification number)
	23–24	I2	*NLIST1 (Index governing printout of wind rose input data: if NLIST1 $\leq$ 0, data is printed)
	25–26	I2	*NLIST2 (Index governing printout of source input data: if NLIST 2 $\leq$ 0, data is printed)
	27–31	I5	IRD (Data input file number)
	32–36	I5	IWR (Output print file number)
	37–41	I5	IPU (Output punch file number)
	42–59	2F9.0	CA(1)–CA(2) (Constants of the linear equation $Y = CA + CB \times X$, used to calibrate the calculated concentrations of the two pollutants considered in the model)
	60–77	2F9.0	CB(1)–CB(2) (Slope of the linear equation $Y = CA + CB \times X$ used to calibrate the calculated concentrations of the two pollutants considered in the model)
3	1–6	I6	*NCALR (Total number of receptors which will be used in computing calibration coefficients for either or both pollutants. Leave blank if coefficients not to be computed. NCALR $\leq$ 50)
	8–11	I4	*IOCAL (Indicates calibration option desired: If IOCAL = 0, regression constants are input and not computed. If IOCAL = 1, constants will be computed and processing will stop if confidence level not satisfactory. Otherwise, constants will be used to calibrate.

[a]Asterisks denote additions to or changes in card input sequence given in Table 6 of the CDM User's Guide.

[b]The data listed on "cards" 1 and 2 must in fact be input on cards. All data on subsequent "cards" must be input from logical unit number IRD, provided on card number 2. This unit may be the card reader or any other input device which can supply the data in card image format.

Table 4-2. Card Input to CDMQC[a,b] (cont.)

Card No.	Column	Format	Contents
3 (contd.)	8-11	I4	If IOCAL = 2, constants will be computed and default values (slope = 1, intercept = 0) will be used to calibrate if confidence level not satisfactory. Otherwise, calculated constants will be used to calibrate. If IOCAL = 3, constants will be computed, the results printed and processing will stop.)
	13-22	F10.0	*BKGR(1) (Arithmetic mean background concentration of pollutant 1, in micrograms/cubic meter)
	23-32	F10.0	*BKGR(2) (Arithmetic mean background concentration of pollutant 2, in micrograms/cubic meter)
	34-37	A4	*LPNAM(1) (Name of pollutant 1)
	39-42	A4	*LPNAM(2) (Name of pollutant 2)
	44-55	3F4.0	*PAV(1,1)-PAV(3,1) (Up to three desired averaging times (hours) for statistical output for pollutant 1)
	57-68	3F4.0	*PAV(1,2)-PAV(3,2) (Up to three desired averaging times (hours) for statistical output for pollutant 2)
	70-74	F5.0	*CTOF (Percentage: sources contributing less than this percent to total calibrated concentration will not be individually listed in any culpability lists)
4	1-6	F6.0	DELR (Initial integration increment of radial distance from receptor, meters)
	7-12	F6.0	RAT (Ratio of length of a basic emission grid square and the length of a map grid square)
	13-18	F6.0	CV (Conversion factor which upon multiplication by RAT expresses the distance of the side of an emission grid square in meters. For example, if the map units are in kilometers, CV = 1000.)
	19-24	F6.0	HT (Average afternoon mixing height in meters)
	25-30	F6.0	HMIN (Average nocturnal mixing height in meters)
	31-36	F6.0	XG (X map coordinate of the southwest corner of the emission grid array)
	37-42	F6.0	YG (Y map coordinate of the southwest corner of the emission grid array)
	43-48	F6.0	XGG (X map coordinate of the southwest corner of the plotting grid)

Table 4-2. Card Input to CDMQC[a,b] (cont.)

Card No.	Column	Format	Contents
4 (contd.)	49–54	F6.0	YGG (Y map coordinate of the southwest corner of the plotting grid)
	55–60	F6.0	RATG (Ratio of the length of the grid square used for plotting and the length of a map grid square)
	61–66	F6.0	TOA (Mean atmospheric temperature in degrees centigrade)
	67–72	F6.0	TXX (Width of basic emission square in meters)
5	1–6	F6.0	DINT (Number of intervals used to integrate over a 22.5° sector. Maximum value is 20; typical value is 4.)
	7–12	F6.0	YD (Ratio of average daytime emission rate to the 24 hour emission rate average.)
	13–18	F6.0	YN (Ratio of the average nighttime emission rate to the 24 hour emission rate average)
	19–54	6F6.0	SZA (1)–SZA (6) (Initial σ_z in meters for each stability class. Six different values can be used, but normally only one value is used.)
	55–66	2F6.0	GB(1)–GB(2) (Decay half life in hours for the two pollutants)
6–101	1–49	[7X, 6F7.0]	F(i,j,k) (Joint frequency function, identical to $\phi(k,l,m)$; i = index for stability class, j = index for wind speed, k = index for wind direction)
[Point source cards follow]			
102[c]	1–6	F6.0	X (X map coordinate of a point source)
	7–13	F7.0	Y (Y map coordinate of a point source)
	*ᵃ	*	*
	21–36	2F8.0	S1-S2 (Source emission rate in grams per second for the two pollutants)
	37–43	F7.0	SH (Stack height in meters)
	44–48	F5.0	D (Diameter of stack in meters)
	49–55	F7.0	VS (Exit speed of pollutants from stack in meters per second)
	56–62	F7.0	T (Gas temperature of stack gases in degrees centigrade)

[c]There will be as many cards of this type as there are point sources. The maximum number of point sources which can be handled is 200. The next card type will arbitrarily be numbered 300.

Table 4-2. Card Input to CDMQC[a,b] (cont.)

Card No.	Column	Format	Contents
102 (contd.)	63–67	F5.0	SA (If this field is blank, Briggs' formula is used to compute stack height. Otherwise, the product of plume rise and wind speed is entered in square meters per second)
300	—	—	*This is a blank card which follows information on the emission point sources. It is used to test the end of the point sources and must not be left out.
[Area source cards follow]			
301[d]	1–6	F6.0	X (X map coordinate of the southwest corner of an area emission grid square)
	7–13	F7.0	Y (Y map coordinate of the southwest corner of an area emission grid square)
	14–20	F7.0	TX (Width of an area grid square in meters)
	21–36	2F8.0	S1-S2 (Source emission rate in grams per second for the two pollutants)
	37–43	F7.0	SH (Stack height in meters)
1000	—	—	This is a blank card which follows information on the area emission sources. It is used to test the end of sources and must not be left out.
[Receptor cards follow]			
1001[e]	1–8	F8.0	RX (X map coordinate of the receptor)
	9–16	F8.0	RY (Y map coordinate of the receptor)
	30–35	F6.0	*COBS(1)[f] (Measured concentration of the first pollutant at the receptor in micrograms/cubic meter. Leave blank if not known.)
	36–41	F6.0	*COBS(2)[g] (Measured concentration of the second pollutant at the receptor in micrograms/cubic meter. Leave blank if not known.)
	43–47	F5.0	*SGD(1)[h] (Standard Geometric Deviation (24 hour) of pollutant 1 to be used for output at other averaging times)

[d]There will be as many cards of this type as there are area sources. The maximum number of area sources which can be handled is 2500. The next card type will arbitrarily be numbered 1000.

[e]There will be as many cards of this type as there are receptors. The maximum number of receptors which may be handled is 200.

[f]Required only if this receptor is to be used in calibration for pollutant 1.

[g]Required only if this receptor is to be used in calibration for pollutant 2.

[h]Required only if Larsen statistical output is desired for pollutant 1 at this receptor.

Table 4-2. Card Input to CDMQC[a,b] (cont.)

Card No.	Column	Format	Contents
1001 (contd.)	48–52	F5.0	*SGD(2)[i] (Same as SGD(1), but for pollutant 2)
	54–55	I2	*IPNCH (A control parameter which, if greater than zero will cause standard concentration output to be punched.)
	56–57	I2	*NROSE (A control parameter for concentration rose output: If NROSE = 0 (blank), no concentration rose data will be printed or punched; If NROSE = 1, concentration roses will be printed but not punched; If NROSE = 2, concentration roses will be printed and punched.)
	58–59	I2	*NCULP (A control parameter which specifies source contribution list option [print only]: If NCULP = 0, no list is printed; If NCULP = 1, list for pollutant 1; If NCULP = 2, list for pollutant 2; If NCULP = 3, list for both pollutants.)
	60–61	I2	*NLARS (A control parameter which specifies Larsen statistical output option [print only]: If NLARS = 0 [blank], no statistical output; If NLARS = 1, for pollutant 1 only; If NLARS = 2, for pollutant 2 only; If NLARS = 3, for both pollutants.)
	77–80	A4	*NRAM (Optional receptor identification name)

[i]Required only if Larsen statistical output is desired for pollutant 2 at this receptor.

Source: USEPA, 1977

collected. Fortunately, in the U.S., the data are available from the regulatory agencies for which the impact analyses are being prepared. Unfortunately, the emissions database does not appear to be updated frequently enough so that in many areas the data available are quite out of date. If the local situation has changed dramatically, i.e. a new industry has located there or an old one has gone out of business, this can have a major impact on model results.

The eighth set of cards is a similar inventory for area sources of air pollution. The major problem with these data is that estimates are relatively difficult to obtain. In most cases, one would have to "borrow" emission rates from other studies based on the similarity of land uses between the study area being simulated and others found in the literature. One of the other models reviewed below, the LAN-

Table 4-3. Outputs of CDMQC Air Pollution Model

Item	Comments
Pollutants Modeled	Up to 2
Geographic Grid	
Calculation Parameters	
Meteorological Assumptions	
Background Concentrations	
Meteorological Joint Frequency Function	Six stability classes, 16 directions, 6 wind speeds
Area and Point Source Inventory	Lists coordinates, emission rates, etc., for each
Calibration Procedure Results	Listed only if options used
Receptor Data	See Table 4-10
Receptor Summary	See Table 4-6
Averaging Time Calculations	See Table 4-8

TRAN model which is part of the AQUIP package, more explicitly estimates these area sources. In the CDM model, the user must define all the parameters either through correlation with the literature or through some actual field sampling of emissions.

The final set of input data are the locations of the pollutant receptors. Along with the locations of the receptors, one can include the observed concentrations of the pollutants at the receptor for comparison with modeled values.

Output from the CDMQC model allows one to undertake a number of analyses. The nature of the output is summarized in Table 4-3. First, the output shows the values of the default and control parameters used in the run. Next, the output shows the stability wind rose frequency distribution. If the calibration procedure is used, the results are shown in the next part of the output. For each receptor, the model next gives the contribution of each point and area wind rose direction to the air quality at that location. Then, one can see that the contribution of each of the point and area sources to the pollutant concentration at each receptor is indicated. Following the detail for each receptor comes a summary table showing the results for all receptors, for both pollutants, broken down by point, area, and background sources. The model also provides for the calculation of 24, 48, or 72 hour concentrations of each pollutant based on a calculating formula derived by Larsen (see Larsen, 1974). Using all of these data, one can compare the contribution of new and old sources of pollution, simulate the impact of new sources on ambient levels, and, using multiple runs, simulate the alternative impacts of different locations and plant designs of new sources on ambient levels. An example later in the chapter further illustrates these uses of the model.

The UNAMAP Package

The CDM model is part of a larger group of air pollution models that USEPA has put together in a package of models called UNAMAP. Although we will not review each of these models separately, it might be useful to review the general contents of the package. A computer tape containing all of the models is available from the National Technical Information Service under number PB240-273. A newer version containing several additional models has just become available and is cited in Appendix B. Instructions on ordering from this agency are also given in that section. The contents of this package (the older version) are as follows:

APRAC—an urban carbon monoxide model developed at Stanford Research Institute. This model computes hourly average CO using an extensive traffic inventory as input. Documentation is available under PB213-091.

HIWAY—an interactive program to compute the hourly concentrations of non-reactive pollutants downwind of roadways in areas where one can assume uniform wind conditions and level terrain. A more extensive model overview is given in the following section. Documentation is available under PB239-944/AS.

CDM—the long term, multiple source air pollution model discussed above.

Three point source models: PTMAX; PTDIS; PTMTP—these are all interactive programs that estimate different aspects of point source air pollution.

Many other air pollution models are available from federal sources. Several others are reviewed below. A complete bibliogaphy of the models is given in Appendix B.

HIWAY. The HIWAY model is one of several models that uses a steady state Gaussian approach to the modelling of air pollution from line sources of pollution. The model requires the user to perform several calculations exogenous to the model to use as input data. Specifically, the user must calculate for each lane of traffic, a uniform emission rate. This can be derived using the following formula:

$$q_i(g/\sec/m) = \frac{EF\ (g/veh/mi)\ TV\ (veh/hr)}{1609.3\ (m/mi)\ 3600\ (sec/hr)}$$

$$= 1.726 \times 10^{-7}(EF)(TV)$$

where

q_i = the line source emission rate;
EF = the emission factor;
TV = the traffic volume.

The calculation can be made using the EPA's *Compilation of Air Pollutant Emission Factors* (EPA, 1976) or using the MOBILE1 model discussed below. Emission factors vary with vehicle speed, the number of cold starts, and the age of the vehicle. Thus, the model is quite heavily data-demanding, requiring an estimate of vehicle mix, traffic flow, and traffic speed. Given these inputs, one calculates the emission rate which is then dispersed downwind of the highway using the HIWAY model.

The model uses the input data to calculate the pollutant concentration at selected output receptors. Only one wind condition is used in each run of the model rather than the weighted frequency function that is used in the CDMQC model. Thus, the user must simulate multiple wind and stability conditions in order to determine if an air pollution problem might exist.

Table 4-4 shows the input data requirements for the model. As one

Table 4-4. Input Data Cards—HIWAY

Name	Columns	Format	Form	Variable	Units
Card type 1 (1 card)					
Head	1–80	20A4	AAAA	Alphanumeric data for heading	—
Card type 2 (1 card)					
REP1	1–10	F10.0	XXXX.XXX	East coordinate, point 1	Map units
SEP1	11–20	F10.0	XXXX.XXX	North coordinate, point 1	Map units
REP2	21–30	F10.0	XXXX.XXX	East coordinate, point 2	Map units
SEP2	31–40	F10.0	XXXX.XXX	North coordinate, point 2	Map units
H	41–50	F10.0	XX.X	Height of line source	Meters
WIDTH	51–60	F10.0	XX.	Total width of highway	Meters
CNTR	61–70	F10.0	XX.	Width of center strip	Meters
XNL	71–80	F10.0	X.	Number of traffic lanes	—
Card type 3 (up to 3 cards)					
QLS	1–80	F10.0	.XXXXXXXXX	Emission rate for each lane	g sec^{-1}m^{-1}
Card type 4 (1 card; can be blank for at grade)					
CUT	1–10	F10.0	X.	1, if cut; 0, if at grade	—
WIDTC	11–20	F10.0	XX.	Width of top of cut section	Meters
Card type 5 (1 card)					
THETA	1–10	F10.0	XXX.	Wind direction	Degrees
U	11–20	F10.0	XX.X	Wind speed	π sec^{-1}
HL	21–30	F10.0	XXXX.	Height of mixing layer	Meters
XKST	31–40	F10.0	X.	Pasquill stability class	—
Card type 6 (1 card)					
GS	1–10	F10.0	X.	Scale factor[a]	—
Card type 7 (any number of cards)					
XXRR	1–10	F10.0	XXXX.XXX	East coordinate of receptor[b]	Map units
XXSR	11–20	F10.0	XXXX.XXX	North coordinate of receptor	Map units
Z	21–30	F10.0	XX.	Height (above ground) of receptor	Meters

[a]The scale factor converts map units to kilometers.
  If map units in kilometers, scale factor = 1.0
  If map units in meters, scale factor = 0.001
  If map units in feet, scale factor = 0.000305
  If map units in miles, scale factor = 1.61

[b]To begin again with another set of data, a value of 9999. is punched for XXRR (card type 7) following the last receptor card.

Source: USEPA, 1975.

can see the requirements are much less extensive than for the CDMQC model. This is somewhat misleading however in that so much of the calculation and data handling must be carried out externally to the model.

MOBILE1

Because the emission calculations required for models such as HIWAY are quite cumbersome, another model has been developed that prepares the emission rates for further analysis. This model, MOBILE1, uses data on vehicle speed and mix, the year, percent of hot and cold starts, the temperature, and the presence or absence of an inspection program to predict the weighted emission factors for each of the highways under consideration. The model has imbedded within it the EPA emission factors by model year from the present to the year 2000. Thus the user may easily incorporate all of the variables having an impact on the emission factors. In order to simulate the impact of such emissions on air pollution levels, this model must be used in tandem with HIWAY or with one of the other impact models discussed below.

The one trap that the user can fall into when using this model involves the imbedded emission factors. These are based on EPA's scheduling of air emission requirements for the auto industry. In recent years, this schedule has been rolled back in an effort to aid the ailing U.S. auto industry. In addition, the rate of replacement of the auto fleet is proceeding more slowly than previous estimates have indicated, due to the high cost of new automobiles. Thus, one might get an unrealistic estimate of emissions if one blindly uses the factors incorporated into the model. This, in turn, would produce a rosy estimate of future air pollution problems. As one analyst has put it, "I cannot load the system enough to produce a future air pollution problem" (Charles Gebhardt, 1983). When one couples the MOBILE1 model with an impact model, one must also utilize local monitoring data to validate the model estimates in order to avoid this problem. Future forecasts must be adjusted to reflect the actual vehicle age mix and emission factors in force at that time.

CALINE-2 and CALINE-3

Two other models which allow the simulation of air pollution impacts from line sources were originally built for the California Department of Transportation under the auspices of the Federal Highway Administration (FHWA). These models, called CALINE-2 and CALINE-3, allow for the calculation of CO levels at different levels of detail. The CALINE-2 model, uses data on the vehicle speed, direction, and volume, the type of traffic (freeway versus arterial), its location (CBD versus suburbs), windspeed and direction, and stability class to estimate the carbon monoxide levels on each link of the major highway system. CALINE-3 provides the same sort of output but requires the user to specify much more explicit volume, geometry, and emission data for each link and allows for both straight line and curved roadway segments, and for multiple sources and receptors. The output is then a much more detailed assessment of the CO concentrations at the receptor points. CALINE-2 is most appropriate for a more general "first cut" type of analysis, while CALINE-3 will provide the details needed for more site-specific analyses. Here again, one must exercise caution in using these models in order to avoid the traps indicated above in the discussion of the MOBILE1 model.

The Aquip Model

One problem with the air pollution models reviewed thus far that should be very apparent is that none alone are adequate for assessing the overall, ambient pollution impacts of the combination of point, line, and area sources of air pollution. There is one model that attempts to put all of these aspects of air pollution together. It is called the AQUIP system and was assembled for the Hackensack Meadowlands Air Pollution Study (Reifenstein et al. 1974). In structure, the model is very much like the Storm Water Management Model reviewed in Chapter 3 in that it consists of several major program modules which pass data to each other as output datasets to cards, disk or tape. The system is represented diagramatically in Figure 4-6. Two sets of input data required for the models are exactly analogous to the information required for point and line source models, namely a point source emission inventory and traffic levels with their related

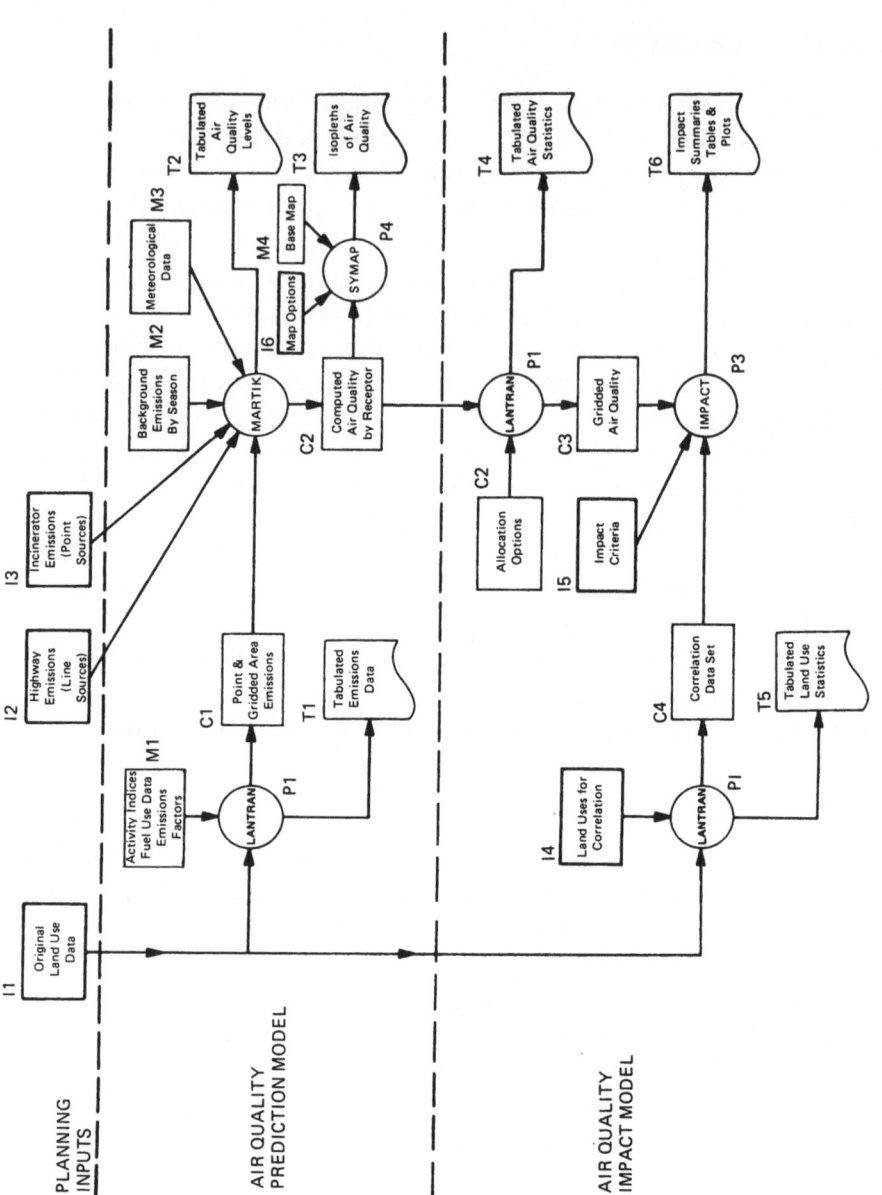

Figure 4-6. AQUIP system. (Source: Reifenstein, 1974.)

emission factors. These are shown as parts I2 and I3 on Figure 4.6. The third major data input is a somewhat unique approach to the modelling of area source emissions. Rather than using sampling techniques or literature estimates of area source emissions, the AQUIP system uses land use activity data as input to computations for the area source emissions (segment I1). The use of land use data make it theoretically possible to incorporate the impacts of land use decisions into the air pollution modelling process. However, data problems for area source emission factors still make this process somewhat subjective.

Given the input data, the AQUIP system uses two computer models to delineate the air quality impacts of the land use activities. First, the LANTRAN model (P1) is used to compute the air quality implications of the land use pattern input to the model. The land use activities are multiplied by emission factors relating to the intensity of use and fuel use and then translated to a rectangular grid system for use by the next model. The second computer model, MARTIK, uses a Gaussian plume model that adds together the impacts of the point, line, and area sources, disperses them, and calculates the combined impacts on ambient air quality at user defined receptors (T2 on Figure 4.6). What distinguishes this model from CDM is the ability to incorporate all *three* sources of air pollution into *one* set of calculations.

There are two other possible sets of calculations that one can undertake using the AQUIP system. First, one can use the SYMAP system to produce an isopleth map of the final air quality (P4 and T3). This requires the digitization of a base map, specification of map options, and linkage with the output of the MARTIK model. Finally, the model output can be compared to a set of standards and also plotted, allowing the user to create illustrations showing those locations where an air quality problem might arise (P3 and T6 on Figure 4.6).

The AQUIP system obviously allows the user to undertake a much more complex set of air pollution computations than is the case with the other models. This complexity, however, brings with it complexity of operation. The methods employed are, for the most part, parallel to those used in the other air pollution models discussed above and therefore are comparable in accuracy.

The exception to this advantage is the calculation of area source contributions. One major problem arises with these computations. They are based on a set of generalized emission factors by land use relating activity level and fuel use to air pollution emissions. Actual data on such activity levels and the types of fuel used by a particular business are not available from secondary sources. Thus the user has only two options, either using the default estimates employed by the AQUIP model builders or gathering this extensive data base by survey. Obviously, the first option can result in major errors of estimation due to the inaccuracy of the input data while the latter method will be quite costly. AQUIP does, then, provide a better method for estimating area source pollution, but to undertake this analysis the user must be willing to make a major data gathering investment.

MODEL VALIDATION

Only a few studies have been made that have attempted to validate empirically the results of air pollution modeling efforts. Several such studies are indicated in the references at the end of the chapter. The findings among these evaluations are very similar to those of Turner et al. (in USEPA, 1973). These authors compared the results of six dispersion models for point and area source emissions with emission data for the New York Air Quality Control Region. They found excellent correlations of the observed with the calculated particulate and sulfur dioxide annual average concentrations. For sulfur dioxide, linear correlation coefficients ranged from 0.77 to 0.89 while for particulates they ranged from 0.57 to 0.66. For several models, the predicted mean values for sulfur dioxide at 75 locations are close to the mean measurements while the root mean square error is less than the standard deviation of the measured values.

Particulate predictions are less accurate with only one model having a prediction close to the actual mean and only one having a root mean square error less than the standard deviation. Some of these errors may be due to the difficulty of obtaining accurate monitoring data for particulates. These results can be interpreted to indicate some systematic biases in the models, the nature of which vary with the particular model and the type of calculations being made. However, it is difficult to draw exact conclusions in this regard because the monitoring data themselves may incorporate sampling biases that could equally be responsible for the errors. Model users should become familiar with any implied biases before employing the models for predictive purposes.

Other, more fundamental questions concerning the models still remain. The nature of the dispersion taking place may not be Gaussian in nature. This is particularly the case in areas which have variable terrain or complex urban structure where microclimatic temperature and wind effects may occur. Only specialized models can begin to address these circumstances. The data input to the various models also remains problematic. One study has shown, for example, that the use of a joint probability distribution for wind speed/direction and stability class can have a marked impact on model results (Guldmann, 1983). The transferability of major airport weather data to disparate areas away from the airport may also be questionable. The required data are usually only available for airport locations. In rural areas away from major airports, the weather conditions may be significantly different and therefore produce another set of errors. Finally, the point, line, and area emissions inventories are not always updated to the most recent time period. Not only might these data be incorrect relative to the location and amount of pollution from each source, but also there may be sources that are intermittent rather than constant, causing further modelling complications.

As is the case with the other models we have discussed, these unanswered technical questions do not prevent the wide use of the Gaussian plume and other models to help answer the environmental questions concerning air pollution. The wise user should keep all of these potential problems in mind to minimize, as much as is feasible, major computational errors when using any of these models.

MODELLING EXAMPLES

Our first example will make use of the CDMQC model to answer some impact questions with regard to a new point source of air pollution. Assume that a power company, the Noveau Power Company, wishes to install a new, 1000MW coal fired power plant in the area west of Dayton, Ohio. You obtain the following information:

1. A point source inventory for the region in question with data on 90 point sources of air pollution from sulfur dioxide and particulates.
2. A joint frequency distribution for weather conditions at the Dayton airport.
3. The background concentrations of the two pollutants.
4. The grid coordinate location for the proposed plant.
5. These data on potential emissions:

> Emission rate for sulfur dioxide—1226 g/sec*
> Emission rate for particulates — 113 g/sec
> Stack temperature — 410.8 degrees K
> Stack diameter — 9.1 meters
> Stack height — 201 meters

With the use of these data, the CDMQC model is employed to simulate the air pollution emissions with and without the proposed new plant and give the air pollutant concentrations at 17 receptor points. The output is summarized in Tables 4-5 through 4-10. Table 4-5 shows the area and point source inventory for the region under consideration, including the proposed plant as point source 91. Table 4-6 is a summary of the SO_2 and TSP levels without the new pollution source at the 17 receptor sites around the region. A similar table showing the pollutant concentrations at the same receptors with the new point source is given as Table 4-7.

By comparing these two simulations, one can see the impacts the new plant has on both of the pollutants being evaluated. First, the impact for TSP is almost nonexistent with the new source not affecting the final TSP levels. The impacts for SO_2 are higher but still not very dramatic for this case. The impacts relative to the 24 hour averages are also not very dramatic. This is shown in Tables 4-8 and 4-9. Here, however, the total concentrations are more significant in that the 24 hour standard for SO_2 is already being violated at receptor 3 (206 ug/m^3 vs. the 180 ug/m^3 standard) and is only made worse by the new plant. Thus, the addition of the plant would probably be allowed only if an appropriate offset was obtained.

Table 4-5. Air Pollution Source Inventory

CLIMATCLCGICAL DISPERSICN MODEL

JCINT FREC. CIST 1 – CAYTCN AREA
RUN 100

AREA AND PCINT SOURCE INVENTCRY

POINT SCURCE ID	X MAP COORDINATE	Y MAP CCCRDINATE	EMISSICN RATE FCR TSP (GRAMS/SEC)	EMISSICN RATE FCR SC2 (GRAMS/SEC)	STACK HEIGHT (METERS)	STACK DIAM (METERS)	STACK EXIT SPEED (METERS/SEC)	STACK GAS TEMP (DEG CELS)	OPTIONAL PLUME RISE CCEFFICIENT (SC METERS/SEC)
1P	241.40	4411.1C	22.66	0.0	20.70	C.50	27.32	65.60	0.0
2P	241.40	4411.80	158.49	2C.72	58.50	4.40	8.89	176.7C	0.0
3P	241.40	4411.8C	21.22	14.76	58.50	4.40	0.C6	176.70	0.0
4P	241.40	4411.80	33.61	23.39	58.50	4.40	0.C6	176.70	0.0
5P	241.40	4411.80	26.12	19.97	58.50	4.40	8.89	176.70	0.0
6P	241.40	4411.8C	5.15	38.40	55.80	5.00	2.37	122.80	0.0
7P	252.00	4409.0C	C.86	1.15	38.10	1.70	0.10	276.70	0.0
8P	243.20	4413.60	195.37	20.39	24.40	5.C0	2.14	148.90	0.0
9P	243.10	4413.00	2.61	57.9C	25.90	1.80	1.79	132.20	0.0
10P	243.10	4413.00	2.59	0.0	25.90	1.80	1.79	132.20	0.0
11P	235.00	4409.4C	3.65	0.72	19.20	1.80	1.52	248.9C	0.0
12P	235.00	4409.4C	3.76	C.74	18.90	2.40	2.44	182.20	0.0
13P	235.0C	4409.40	7.82	1.92	18.90	2.40	2.44	237.80	0.0
14P	235.00	4409.40	11.4C	2.23	18.90	2.40	2.44	237.80	0.0
15P	235.00	4409.40	17.62	4.32	17.40	1.80	0.46	232.80	0.0
16P	235.00	4409.40	25.81	5.84	17.40	1.80	0.46	197.20	0.0
17P	235.00	4409.40	2.57	1.72	19.80	1.80	0.81	175.00	0.0
18P	239.40	4411.30	1.C1	1.CC	8.50	1.10	1.06	282.20	0.0
19P	240.10	4411.80	3.6C	3.57	25.90	1.40	1.21	279.40	0.0
20P	240.10	4411.8C	2.95	2.92	25.90	1.40	1.68	279.40	0.0
21P	239.50	4409.60	1.82	1.26	12.20	1.20	0.61	271.10	0.0
22P	235.50	4409.6C	1.75	1.21	12.20	1.20	0.61	271.1C	0.0
23P	239.50	4409.60	1.45	1.01	12.20	1.20	0.61	271.10	0.0
24P	238.50	4409.60	1.24	C.86	12.30	1.20	0.61	271.10	0.0
25P	238.30	4409.30	1.17	1.15	20.70	1.20	0.46	260.00	0.0
26P	240.60	4410.9C	11.71	3.12	9.10	1.20	0.76	271.10	0.0
27P	267.00	4401.1C	4.4C	C.C	6.10	0.10	0.10	20.00	0.0
28P	246.00	4408.00	2.63	4.88	37.50	3.20	0.13	287.80	0.0
29P	222.90	4404.4C	1.42	0.0	11.60	1.70	6.60	23.90	0.0
30P	222.90	4404.4C	C.8C	C.C	1C.70	2.80	6.46	23.90	0.0
31P	227.50	4405.90	2.15	5.67	31.70	2.40	11.12	232.20	0.0
32P	227.50	4405.9C	2.22	9.09	31.70	2.40	11.12	232.20	0.0
33P	226.30	4405.10	1.63	1.54	48.80	2.40	11.15	232.20	0.0
34P	226.30	4405.10	1.62	4.64	48.80	3.C0	14.75	190.60	0.0
35P	226.30	4405.10	2.76	22.57	48.80	3.00	14.75	190.60	0.0
36P	227.20	4408.10	C.75	C.C	12.20	1.30	1.22	35.0C	0.0
37P	224.00	4405.40	27.35	0.0	9.1C	1.80	4.26	57.20	0.0
38P	223.80	4402.00	1.13	0.0	14.60	6.10	1.54	46.1C	0.0
39P	226.80	4403.1C	C.76	C.C	9.80	0.60	138.81	37.80	0.0
40P	229.6C	4403.10	1.71	1.62	53.30	2.C0	3.98	287.80	0.0
41P	229.60	4403.1C	3.16	2.99	53.30	2.C0	3.98	287.60	0.0
42P	228.40	4412.1C	C.87	0.0	18.60	0.80	10.35	93.30	0.0
43P	228.40	4412.1C	C.64	0.0	0.90	0.90	21.46	40.60	0.0
44P	228.40	4412.1C	2.56	0.0	15.80	1.10	15.98	65.60	0.0
45P	229.10	4407.00	1.9C	C.C	1C.7C	0.70	18.72	20.00	0.0
46P	229.10	4407.00	C.78	0.0	10.70	0.90	3.16	21.10	0.0

Table 4-5. Air Pollution Source Inventory (cont.)

CLIMATOLOGICAL DISPERSION MODEL

JOINT FREQ. DIST 1 - DAYTON AREA   RUN 100

AREA AND POINT SOURCE INVENTORY

POINT SOURCE ID	X MAP COORDINATE	Y MAP COORDINATE	EMISSION RATE (FOR TSP) (GRAMS/SEC)	EMISSION RATE (FOR SO2) (GRAMS/SEC)	STACK HEIGHT (METERS)	STACK DIAM (METERS)	STACK EXIT SPEED (METERS/SEC)	STACK TEMP (DEG CELS)	OPTIONAL PLUME RISE COEFFICIENT (SQ METERS/SEC)
47P	221.10	4392.50	3.74	2.24	9.40	0.30	7.41	232.20	0.0
48P	221.10	4392.50	3.74	2.24	10.40	0.30	7.41	232.20	0.0
49P	221.90	4392.00	5.05	4.63	100.00	4.90	12.64	210.00	0.0
50P	224.90	4402.40	15.08	44.87	93.60	4.90	12.64	210.00	0.0
51P	224.90	4402.40	15.48	131.34	93.60	4.90	12.64	210.00	0.0
52P	224.90	4402.40	17.24	4.71	93.60	4.90	12.64	210.00	0.0
53P	224.90	4402.40	6.71	4.24	93.00	4.90	12.64	210.00	0.0
54P	224.90	4401.20	73.87	13.34	51.70	4.90	12.64	132.60	0.0
55P	221.20	4396.10	2.07	1.00	57.00	3.10	4.36	76.70	0.0
56P	220.70	4414.10	2.87	0.00	57.00	1.80	3.03	182.00	0.0
57P	220.70	4414.40	1.14	3.24	6.70	1.00	5.04	182.00	0.0
58P	228.00	4396.50	7.19	3.16	57.00	1.80	4.04	76.70	0.0
59P	224.00	4388.50	6.29	70.17	10.70	2.40	0.41	204.40	0.0
60P	221.30	4388.40	1.91	52.49	76.20	4.20	15.41	151.70	0.0
61P	217.30	4394.00	0.22	6.25	76.20	4.20	15.41	151.70	0.0
62P	224.00	4394.00	0.00	2.75	204.20	1.20	6.07	260.00	0.0
63P	224.00	4413.60	0.00	1.00	38.10	2.10	2.14	276.70	0.0
64P	225.10	4405.00	0.00	10.20	24.40	5.00	2.14	148.90	0.0
65P	224.10	4405.70	0.00	10.20	24.40	5.00	2.14	148.90	0.0
66P	224.60	4404.70	0.00	4.35	21.30	1.70	14.55	204.40	0.0
67P	225.60	4358.50	0.00	1.33	18.30	1.60	16.55	204.40	0.0
68P	229.90	4398.50	0.00	1.21	18.00	1.20	12.94	233.30	0.0
69P	222.50	4398.50	0.00	4.02	53.30	2.70	12.94	162.80	0.0
70P	223.60	4356.30	0.00	4.93	53.30	5.00	2.07	162.80	0.0
71P	223.60	4356.30	0.00	4.48	61.00	5.00	6.07	176.70	0.0
72P	222.30	4396.10	0.00	4.48	61.00	1.70	3.87	176.70	0.0
73P	209.70	4388.40	0.00	2.71	30.00	1.00	3.78	176.70	0.0
74P	217.30	4388.40	0.00	3.77	67.00	3.00	9.81	182.00	0.0
75P	217.30	4388.40	0.00	14.77	76.20	4.20	6.36	182.00	0.0
76P	216.60	4398.40	0.00	26.11	76.20	4.20	15.94	153.30	0.0
77P			0.00	8.40	61.00	4.30	15.94	153.30	0.0
78P			113.00	122.33	201.20	9.10	26.61	445.00	0.0
79P							23.20	137.00	0.0

AREA SOURCE ID	X COORD OF SW GRID CORNER	Y COORD OF SW GRID CORNER	WIDTH OF GRID SQUARE (METERS)	EMISSION RATE FOR TSP (GRAMS/SEC)	EMISSION RATE FOR SO2 (GRAMS/SEC)	EMISSION HEIGHT (METERS)
1A	216.00	4410.00	12000.00	0.00	0.00	20.00

Table 4-6. Without New Plant Pollution Levels

C L I M A T O L O G I C A L   D I S P E R S I O N   M O D E L
JOINT FREC. DIST 1 — DAYTON AREA
RUN 100

CALIBRATED CONCENTRATION (MICROGRAMS/CU. METER)

POLLUTANT: TSP

POLLUTANT: SC2

RECEPTOR NO. ID	X COORD	Y COORD	POINT SOURCES	AREA SOURCES	BACKGROUND	TOTAL	POINT SOURCES	AREA SOURCES	BACKGROUND	TOTAL
1	241.50	4412.10	30.1	0.0	60.0	90.1	12.9	0.0	10.0	22.9
2	235.30	4393.10	3.3	0.0	60.0	63.3	30.2	0.0	10.0	30.2
3	235.30	4410.60	73.4	0.0	60.0	134.4	20.8	0.0	10.0	30.5
4	234.00	4391.00	20.0	0.0	60.0	80.0	8.5	0.0	10.0	18.5
5	232.00	4401.00	5.7	0.0	60.0	65.6	3.2	0.0	10.0	21.7
6	229.00	4415.00	5.7	0.0	60.0	65.9	11.4	0.0	10.0	25.0
7	229.00	4359.00	3.8	0.0	60.0	63.8	5.7	0.0	10.0	13.1
8	228.00	4297.00	3.3	0.0	60.0	63.3	5.0	0.0	10.0	13.0
9	224.00	4420.00	10.1	0.0	60.0	70.1	3.1	0.0	10.0	13.6
10	235.00	4417.00	10.5	0.0	60.0	70.5	7.6	0.0	10.0	17.6
11	232.00	4414.00	11.4	0.0	60.0	71.4	6.3	0.0	10.0	16.3
12	232.00	4417.00	16.7	0.0	60.0	76.7	8.3	0.0	10.0	18.3
13	235.00	4413.00	5.7	0.0	60.0	65.7	8.7	0.0	10.0	18.7
14	230.00	4409.00	14.1	0.0	60.0	74.1	3.7	0.0	10.0	12.9
15	224.00	4414.00					8.0	0.0	10.0	18.0
16	229.00									

Table 4-7. Pollution Levels with New Plant

C L I M A T O L O G I C A L   D I S P E R S I O N   M O D E L
JOINT FREC. DIST 1 — DAYTON AREA
RUN 100

CALIBRATED CONCENTRATION (MICROGRAMS/CU. METER)

POLLUTANT: TSP

POLLUTANT: SO2

RECEPTOR NO. ID	X COORD	Y COORD	POINT SOURCES	AREA SOURCES	BACKGROUND	TOTAL	POINT SOURCES	AREA SOURCES	BACKGROUND	TOTAL
1	241.50	4412.10	30.2	0.0	60.0	90.2	13.2	0.0	10.0	23.2
2	235.30	4393.10	3.3	0.0	60.0	63.3	3.7	0.0	10.0	13.7
3	235.30	4410.60	73.4	0.0	60.0	133.4	21.0	0.0	10.0	31.0
4	234.00	4391.00	20.1	0.0	60.0	80.1	8.9	0.0	10.0	18.9
5	232.00	4401.00	5.7	0.0	60.0	65.6	3.6	0.0	10.0	13.6
6	229.00	4415.00	8.7	0.0	60.0	68.7	11.8	0.0	10.0	21.6
7	229.00	4359.00	3.3	0.0	60.0	66.3	6.3	0.0	10.0	16.3
8	228.00	4297.00	3.3	0.0	60.0	60.3	3.4	0.0	10.0	13.4
9	224.00	4420.00	10.1	0.0	60.0	70.1	3.4	0.0	10.0	13.4
10	235.00	4417.00	8.5	0.0	60.0	68.5	7.1	0.0	10.0	17.1
11	232.00	4414.00	16.7	0.0	60.0	76.7	7.1	0.0	10.0	17.1
12	232.00	4413.00			60.0	76.4	8.7	0.0	10.0	18.7
13	235.00	4409.00	3.4	0.0	60.0	63.4	4.20	0.0	10.0	14.20
14	224.00	4414.00	14.1	0.0	60.0	74.1	8.4	0.0	10.0	18.4
15	229.00									

Table 4-8. 24 Hour Concentrations before New Plant

C L I M A T O L O G I C A L   D I S P E R S I O N   M O D E L
JOINT FREQ. DIST 1—DAYTON AREA
RUN 100

STATISTICAL DATA AT SELECTED RECEPTORS (MICROGRAMS/CU. METER)

POLLUTANT: SO2

AVERAGING TIME = 24.0 HOURS

RECEPTOR NUMBER	RECEPTOR ID	X COORD	Y COORD	EXPECTED GEOMETRIC MEAN	EXPECTED MAXIMUM CONCENTRATION	STANDARD GEOMETRIC DEVIATION	EXPECTED ANNUAL ARITHMETIC MEAN
1		241.50	4412.10	17.40	154.06	2.10	22.91
2		243.30	4410.30	10.04	88.95	2.10	13.23
3		235.30	4410.60	23.35	206.82	2.10	30.75
4		234.00	4405.000	14.04	124.22	2.10	18.48
5		236.000	4406.000	10.02	88.70	2.10	13.43
6		207.000	4401.000	11.65	104.85	2.10	13.74
7		218.000	4295.000	9.58	87.21	2.10	12.17
8		214.000	4297.000	7.83	88.35	2.10	13.14
9		223.000	4422.000	9.58	69.18	2.10	10.50
10		232.000	4417.000	13.34	118.18	2.10	17.57
11		228.000	4414.000	12.74	112.87	2.10	16.34
12		232.000	4415.000	13.52	123.74	2.10	18.32
13		220.000	4406.000	10.37	91.82	2.10	13.89
14		224.000	4407.000	10.81	86.90	2.10	13.92
15		229.00	4414.00	13.66	120.99	2.10	17.99

Table 4-9. 24 Hour Concentrations after New Plant

C L I M A T O L O G I C A L   D I S P E R S I O N   M O D E L
JOINT FREQ. DIST 1—DAYTON AREA
RUN 100

STATISTICAL DATA AT SELECTED RECEPTORS (MICROGRAMS/CU. METER)

POLLUTANT: SO2

AVERAGING TIME = 24.0 HOURS

RECEPTOR NUMBER	RECEPTOR ID	X COORD	Y COORD	EXPECTED GEOMETRIC MEAN	EXPECTED MAXIMUM CONCENTRATION	STANDARD GEOMETRIC DEVIATION	EXPECTED ANNUAL ARITHMETIC MEAN
1		241.50	4412.10	17.58	151.74	2.10	23.16
2		243.30	4410.30	10.37	208.20	2.10	13.06
3		234.30	4411.60	24.54	126.78	2.10	31.66
4		234.00	4391.000	16.31	191.32	2.10	18.85
5		236.000	4401.000	16.54	146.50	2.10	13.58
6		207.000	4405.000	10.18	187.91	2.10	21.18
7		218.000	4395.000	10.14	86.81	2.10	13.36
8		214.000	4292.000	17.82	120.65	2.10	13.44
9		223.000	4415.000	13.62	69.81	2.10	10.40
10		232.000	4420.000	13.01	115.24	2.10	17.14
11		228.000	4414.000	14.24	125.81	2.10	18.11
12		224.000	4415.000	10.29	155.56	2.10	14.20
13		224.000	4406.000	10.79	87.71	2.10	14.21
14		225.00	4415.00	13.54	123.50	2.10	18.36

Table 4-10. Detailed Receptor Data

CLIMATOLOGICAL DISPERSION MODEL
JOINT FREQ. DIST 1 - DAYTON AREA
RUN 100

RESULTS: RECEPTOR NUMBER   3 (   )

X COORDINATE: 235.30
Y COORDINATE: 4410.80

PCINT ROSES (MICROGRAMS/CU. METER)

CARD ID	N	NNE	NE	ENE	E	ESE	SE	SSE	S	SSW	SW	WSW	W	WNW	NW	NNW
P1 (SOP)	0.0	0.0	0.0	1.1	2.2	0.6	0.6	0.0	0.0	15.0	1.5	0.6	0.0	0.1	0.1	0.0
P2 (SO2)	0.0	0.0	0.0	1.1	2.2	0.6	0.6	0.0	0.0	15.0	1.5	0.6	0.0	0.1	0.1	0.0

AREA ROSES (MICROGRAMS/CU. METER)

CARD ID	N	NNE	NE	ENE	E	ESE	SE	SSE	S	SSW	SW	WSW	W	WNW	NW	NNW
A1 (TSP)	0.0	0.0	0.0	0.0	0.0	0.0	0.0	0.0	0.0	0.0	0.0	0.0	0.0	0.0	0.0	0.0
A2 (SO2)	0.0	0.0	0.0	0.0	0.0	0.0	0.0	0.0	0.0	0.0	0.0	0.0	0.0	0.0	0.0	0.0

INDIVIDUAL SOURCE CONTRIBUTIONS

	POLLUTANT: TSP		POLLUTANT: SC2	
PCINT SOURCES	MICROGRAMS/ CU. METER	PERCENTAGE OF TOTAL	MICROGRAMS/ CU. METER	PERCENTAGE OF TOTAL

Table 4-10. Detailed Receptor Data (cont.)

CLIMATCLCGICAL CISPERSICN MODEL

JGINT FREC. CIST 1 - CAYTCA AREA
RUN 1CC

RESULTS: RECEPTCR NUMBER 3 ( )

INDIVIDUAL SOURCE CONTRIBUTICNS

	POLLUTANT: TSP		POLLTANT: SC2	
	MICRCGRAMS/ CU. METER	PERCENTAGE CF TCTAL	MICRCGRAMS/ CU. METER	PERCENTAGE CF TOTAL
PCINT SCURCES				
AREA SCURCES				
1A	0.CO	0.00	0.00	0.00

Table 4-11. Columbus "Hot" Spots, CALINE-2

Location	Maximum 8-Hour Concentration (ppm) (STD = 9.0 ppm)
Third Street,	
between Fulton and Mound	12.67
between Mound and Main	11.58
between State and Broad	10.81
between Broad and Gay	9.76
between Gay and Long	9.72
Broad Street,	
between High and Third	10.91
between Fourth and Grant	11.50
between Grant and Cleveland	11.83
between Cleveland and Washington Avenue	11.83
between Washington Avenue and I-71	17.21
between S.R. 315 and Grub	10.88
between Belle and Washington Blvd.	11.48
between Washington Blvd. and Civic Center	9.80
between Civic Center and Front	10.08
Long Street,	
between Front and High	12.66
between High and Third	10.23

Table 4-12. CALINE-3 Output for Columbus
Receptor Locations and Model Results

Receptor	Coordinates (M)			Total (PPM)
	X	Y	Z	
1. Recp. 1	−239	463	0.0	3.8
2. Recp. 2	−506	404	0.0	1.9
3. Recp. 3	35	594	0.0	6.3
4. Recp. 4	415	594	0.0	4.0
5. Recp. 5	−1465	−15	0.0	4.6
6. Recp. 6	−417	15	0.0	2.0
7. Recp. 7	35	18	0.0	2.3
8. Recp. 8	533	18	0.0	1.8
9. Recp. 9	1218	18	0.0	1.8
10. Recp. 10	309	−1028	0.0	7.4
11. Recp. 11	−145	30	0.0	5.3
12. Recp. 12	15	−378	0.0	4.3
13. Recp. 13	15	−974	0.0	6.1
14. Recp. 14	243	30	0.0	6.6
15. Recp. 15	243	−478	0.0	6.0
16. Recp. 16	248	−1051	0.0	6.7
17. Recp. 17	374	30	0.0	4.9
18. Recp. 18	374	−532	0.0	4.5
19. Recp. 19	374	−1084	0.0	5.5
20. Recp. 20	30	−18		5.6

Going on to Table 4-10, we see that the CDM model provides several other valuable pieces of information. The actual percentage contribution by wind rose direction and point source is given for each receptor being modelled. The receptor that is over standards is most critically impacted by point sources 15 and 16 with about 14% and 19% contributions to SO_2 levels while the new source, 91, only contributes 0.80% to this receptors SO_2 level.

Using the full array of output from this model, it is possible to answer many of the policy questions which arise in air pollution control. Noticeable by their absence are data on area sources of pollution (although this model will handle these) and line sources of pollution. The next illustration touches on the omission of this latter emission.

Under the requirements of the Clean Air Act, each community must implement a State Implementation Plan showing how they will meet the required standards. In conformance with this requirement, the Mid-Ohio Regional Planning Commission undertook analysis of CO problems in central Ohio. A two step process was used. First, the CALINE-2 model was used to define hot spots—places where the CO standards seem to be violated—using summary data such as average wind speeds. The results of this analysis are illustrated by Table 4-11 showing the locations of violations of the maximum 8 hour standard for CO. A number of runs of CALINE-3 were then made to determine in more detail if and where the problems were actually included. Data for these runs also included actual monitored windspeed information rather than worst case data. A representative result is shown as Table 4-12. For each of the receptors indicated, none violates the standard (shown by the rightmost column of the table). It should be noted, however, that the modelled violations did occur when worst case wind data were used and only "disappeared" when "real" wind data, with brisker winds, were used in the simulation. Such is the magic of simulation modelling and further shows why actual air pollution monitoring data are critical to help validate such efforts.

REFERENCES

Benson, Paul E. (1979). *CALINE-3—A Versatile Dispersion Model for Predicting Air Pollutant Levels Near Highways and Arterial Streets*. Sacramento: California State Department of Transportation. Springfield, VA: NTIS PB80-220841.

Clean Air Act as Amended, 42 U.S.C. SS7401-7642, formerly 43 U.S.C. SS1857-1858a, 1977.

Federal Highway Administration, (1976). *A User's Manual for CALINE-2 Computer Program* by K. E. Jones et al. Springfield, VA: NTIS PB-271106.

Gebhardt, Charles (1983), Ohio Department of Transportation. Personal communication.

Guldmann, Jean-Michel (1983). "A Structural Framework for the Design of Integrated Environmental and Land-Use Planning Optimization Models." Paper presented at the 23rd European Congress of the Regional Science Association, Poltiers, France, August 30–September 2.

Guthman, Lewis E. (1978). *User's Guide to MOBILE1 Mobile Source Emissions Model.* U.S. Environmental Protection Agency. Springfield, VA: NTIS PB81-159964.

Larsen, R. I. (1974). "A Mathematical Model for Relating Air Quality Measurements to Air Quality Standards." *Office of Air Programs Publications No. AP-89.* Office of Technical Information and Publications, U.S. Environmental Protection Agency. Springfield, VA.: NTIS PB-205277.

Mills, Michael T. et al. (1981). *Evaluation of Point Source Dispersion Models.* Teknekron Research Inc. for USEPA. Springfield, VA: NTIS PB82-121062.

Reifenstein, Edward C. et al. (1974). *Hackensack Meadowlands Air Pollution Study—AQUIP Software System User's Manual.* Environmental Research and Technology, Inc. Springfield, VA: NTIS PB-238605.

Singpurwalla, Nozer D. (1975). "Models in Air Pollution," in *A Guide to Models in Governmental Planning and Operation,* ed. by Saul I. Gass and Roger L. Sisson. Potomac, MD: Sauger Books.

Stern, Arthur C., editor (1976). *Air Pollution,* 3rd Ed. 7 vols. New York: Academic Press.

Turner, D. B., J. R. Zimmerman, and A. D. Busse (1973). "An Evaluation of Some Climatological Dispersion Models." App. E in *User's Guide for the Climatological Dispersion Model.* Springfield, VA: NTIS PB-227 346.

U.S. Air Force, Air Weather Service (1971). *Guide to Local Diffusion of Air Pollutants.* Technical Report 214. Springfield, VA: NTIS AD-726984.

U.S. Environmental Protection Agency (1977a). *Coal Cleaning with Scrubbing for Sulfur Control: An Engineering Economic Summary.* Washington, DC: Decision Series, USEPA, EPA-600/9-77-107.

U.S. Environmental Protection Agency (1976). *Compilation of Air Pollutant Emission Factors, Second Edition.* 2 vols. Springfield, VA: NTIS PB-264194, PB-264195.

U.S. Environmental Protection Agency, (1974). *A Guide for Considering Air Quality in Urban Planning.* Research Triangle Park, NC: EPA-450/3-74-020.

U.S. Environmental Protection Agency (1973). *User's Guide for the Climatological Dispersion Model.* Springfield, VA: NTIS PB-227346.

U.S. Environmental Protection Agency (1977). *Addendum to User's Guide for Climatological Dispersion Model.* Research Triangle Park, NC: EPA-450/3-77-015.

U.S. Environmental Protection Agency (1975). *User's Guide for HIWAY, A Highway Air Pollution Model.* Springfield, VA: NTIS PB-239944/AS.

U.S. Environmental Protection Agency (1978). *User's Guide for PAL.* Research Triangle Park, NC: EPA-600/4-78-013.

U.S. Environmental Protection Agency (1977c). *User's Manual for Single-Source (CRSTER) Model.* Research Triangle Park, NC: EPA-450/2-77-013.

U.S. Environmental Protection Agency (1977b). *Valley Model User's Guide.* Research Triangle Park. NC: EPA-450/2-77-018.

---

*Calculated based on analysis in EPA document (USEPA, 1977a).

5
Land Capability Evaluation

INTRODUCTION

Thus far, we have reviewed models which are very site specific in their focus. Even if they are used for regional evaluations, they must begin with site specific data at a fairly large scale. Model outputs are also very explicit, being predictive and deterministic. Thus, given a particular environmental policy, a particular action can be evaluated to determine if the standards set are going to be met. The subject of this chapter is a set of methods that are entirely different in focus. These methods tend to be regional in orientation, broad in scope, and very subjective at the policy decision points.

They have been given various names including *natural resource evaluation, land capability analysis, suitability analysis,* and *land impact analysis.* The basic idea behind such analyses is to identify those land resources that, if developed, could result in the loss of valuable natural or man-made resources and environmental degradation. Once such an identification is made, the implication is that policies will be implemented to protect the specific resources. In other words, the focus of these analyses is the ultimate "wise" use of our land resources. As we will see, such policies are not always apparent from the analyses undertaken.

An example might help to illustrate how the method works. Let us imagine a township (or other subcounty administrative unit) which is located on the urban-rural fringe. Washington Township might be a good name. The township has a mix of land uses but remains mostly rural and undeveloped with the predominant land uses being agriculture, forest, and open space. The township trustees, realizing that there is increasing pressure for urbanization because of the growth of nearby Metropolitan City, want to do something to protect the community's natural assets. (To some, such a realization at the local level may seem unrealistic, but it is not all that uncommon.) State and Federal officials are worried as well, because Washington Township is part of the important Koochiekoochie River basin and is also a Class I Air Quality Control Region.

Just what resources are there and what land should be protected while other land is developed remains very unclear. Thus, the township trustees, with some state and federal monies, decide to sponsor a natural resources inventory and land capability evaluation. A consultant is hired, who carries out the following procedure:

1. A list is made of all the potentially important variables that could have an impact on the township's resources, for example, slope, forest resources, soil productivity, soil erodibility, landslide potential, adjacent water resources, and historical sites.

2. The importance of each of these variable-resources to the township is determined using a subjective ranking process. The consultant, based on experience, assigns different values of the variables according to their importance. For example, lands within 1000 feet of the riverbank are rated the highest because their improper development could have the maximum adverse consequences on water quality. Land more than 1000 feet away would supposedly pose a lesser problem. Similarly, land with steep slopes, say, greater than 10%, would pose a problem for most development because of erosion and mass wasting problems and high development costs. Land with slopes from 5% to 10% would pose moderate hazards while those with slopes of less than 5% would pose few hazards.

3. With the aid of some site surveys, aerial photography interpretation, and secondary sources of information such as soil surveys and geologic maps, an inventory is made of all the important variables and put onto a set of maps. The maps indicate the location and relative importance of each variable according to its value.

4. The maps are overlaid physically by placing them on plastic sheets and using shading to depict importance, or they are overlaid with computer techniques. These methods will be more completely described below.

5. The most important areas in the township from a resource standpoint are indicated as those places where the most critical variables converge. For example, areas within 1000 feet of the riv-

erbank with greater than 10% slope would be extremely sensitive to development.

6. A report is produced, accompanied by final maps showing the most important areas of the township to preserve. This report to the township trustees is the final product of the consultant.

Several things should now be apparent about land capability evaluation. First, the output of the process is not predictive in the same sense as the other models that have been reviewed. The output instead can be thought of as probabilistic in nature: if the assumptions concerning the nature of the development and its impacts on various resources are correct, the occurrence of such development in sensitive areas will probably have an adverse impact on the local environment. The exact nature of this impact is implicit rather than explicit. For a particular development in a particular location, there may be one or several implicit impacts.

The second aspect of this analytical process that should be apparent is the high degree of subjectivity associated with it. Selection of what to measure and evaluate, and how values will be assigned, is a matter of the judgment of a planning professional. Reasonable people could easily differ in their judgments according to their varying levels and types of experience.

There are many other problems with this process that will be discussed further on. But for now we might conclude our example by pointing out that the alternatives being investigated are in fact critical to the future environmental quality of the area. The problem facing the township trustees upon completion of the analysis is what steps to take to ensure the future environmental well-being of their area without creating an overly burdensome regulatory system. These decisions are particularly difficult in the face of the subjectivity associated with the land capability analysis.

What various approaches to land capability analysis try to accomplish, then, is the establishment of a rational and complete method for determining the potential impacts of land development on the environment and to provide for a planning and regulatory process which will avoid the most adverse impacts and promote those that are the least problematic. The obvious answer to these questions is to build a computer information system which performs all of the required functions. Unfortunately, such an undertaking has not proved effective in all circumstances. Below, we will review the alternative approaches to this noble process, their problems and successes, and then review practical computer approaches to these analyses.

A HISTORY OF CAPABILITY ANALYSIS

A number of individuals are credited with "inventing" and popularizing the idea of capability analysis. There is no doubt that a number of practicing planners, architects, and engineers informally applied these methods well before several individuals wrote down their ideas on the subject. Frequently, three individuals are credited with being the first to employ similar approaches to the problem of capability evaluation: G. Angus Hills, Philip Lewis, and Ian McHarg; all applied one or another form of this analysis for their purposes. An excellent review of this original work is provided in a document published by the Conservation Foundation (Harvard University, Department of Landscape Architecture, 1967). The original purpose of some of this capability analysis was the assessment of the natural environment for purposes of recreation planning, but it was soon expanded to encompass other land uses.

Each of the three individuals mentioned had a slightly different approach to capability analysis. G. Angus Hills, a Canadian forester, based his classification of land resources on physiographic characteristics. The system is an hierarchical one, breaking down the environment into increasingly narrower units based on various differentiating characteristics. For example, the system differentiates units based on macro-climate, landform, geology, soils, and micro-climate. The smallest unit of classification is called the *physiographic site type.* These types are compared to a set of land use categories for determination of suitability. "The units are ranked on a scale of values corresponding to the degree of potential or limitation for each use or activity." (Harvard University Dept. of Landscape Architecture, 1967, p. 8) The highest-intensity use that can be accommodated without site deterioration is determined and the unit is placed into a capability class, from A to G, based on this determination. Table 5-1 shows the capability class definitions. As one can see, the planner must have a particular use or set of uses in mind in order to judge which capability class the physiographic unit might fit. Finally, based on the capability classification, units are assembled into larger geo-

Table 5-1. Definition of Capability Classes A to G on a Comparative Basis

Class	Level of Capability	Relative Intensity of Use Potential[a]	Degree of Limitation	Relative Effort to Obtain and Maintain a High Intensity of Use
A	Very high	Very high	Very low	Not significant
B	High	High	Low	Very low
C	Mod. high	Mod. high	Mod. low	Low
D	Moderate	Moderate	Moderate	Moderate
E	Mod. low	Mod. low	Mod. high	High
F	Low	Low	High	Very high
G	Very low	Very low	Very high	Prohibitory

[a]These are levels to be expected under present normal inputs and social pressures disregarding local differences in present vegetative cover, economic and social factors.

Source: G. A. Hills, *Definition of Capability Classes and Benchmark Sites for the Recreational Land Inventory* as cited in Harvard University, Department of Landscape Architecture, 1967.

graphic areas called *landscape units* whose suitability is again ranked on a scale of 1 to 7.

Philip Lewis, a landscape architect from the University of Wisconsin, uses a landscape analysis incorporating perceptual landscape qualities—interesting views, presence of highways and buildings, vegetation—along with soil survey information to reflect the quality of the land resources. All of these factors are mapped on transparent overlays and combined to reflect the major patterns of land suitable for a particular purpose. For those areas that possess the resources necessary for a particular use, points are assigned reflecting the values of the resources. Points are then totaled to produce specific areas that are "best" for a particular use.

The ecological inventory process of Ian McHarg, a landscape architect and planner from the University of Pennsylvania, incorporates some of the features of both of the other approaches. His approach gathers information on the basic natural and man-made attributes of the area, maps them on transparent overlays, then ranks each land area relative to its suitability for one or more land uses. The rankings, given on a scale of 1 to 7, are represented on overlay maps where the darkest shading indicates the most unsuitable ranking for that resource for the land use under consideration and the lightest shades or colors indicate the most suitable locations. The overlays are then combined to indicate those areas where there is a confluence of the most suitable conditions. The reader is referred to these authors' original publications for further exposition on these methods (Hills, 1966; Lewis, 1964; McHarg, 1966).

Applied Example

A simple example taken from an area in Franklin Township, Ohio, may help to illustrate the usefulness and problems associated with these methods. Although one cannot exactly see how the illustrations would look as overlays, Figures 5-1 to 5-9 represent a typical set of maps that could be used in a capability analysis. Figure 5-1 shows a proposed land use change from rural open space uses to residential development and a small shopping area. The questions which must be answered relate to the potential impacts of several developments on the environment and the potential impacts of the environment on these proposed developments. This undeveloped area is not served by either public sewer or water systems. Thus, these services must be provided on site using wells for water supply and septic systems for sewage disposal.

Figures 5-2 to 5-7 were compiled using a soil survey for Franklin County and show some of the variables critical to the impact analysis. Figure 5-2 indicates the presence of drainage problems in the local soils. Here, one can see that a large portion of the area is subject to slow drainage, ponding, or flooding. Figure 5-3 is another indicator of this drainage problem with soils that have a seasonally high water table. Both of these characteristics would create negative problems for residential development in the form of wet basements, yards that are unusable part of the year, and septic systems which do not function properly. This latter problem carries with it some health risks, namely raw sewage being transmitted in the local groundwater and contaminating the wells.

Some of these soils are also problematic with regard to bearing capacity (Figure 5-4). One area has a poor bearing capacity potentially resulting in problems with roads and building foundations unless extraordinary compensatory building methods are used. The erosion potential (Figure 5-5) of many of the soils are quite high, with

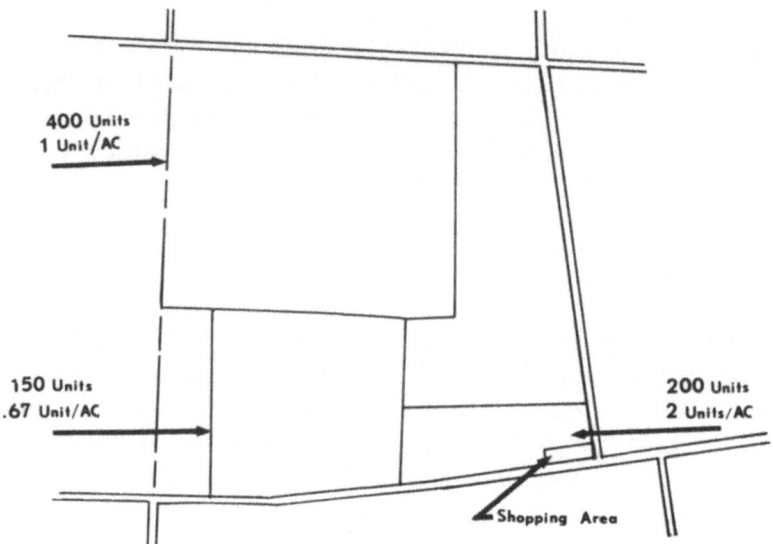

Figure 5-1. Proposed land use: single family dwellings, shopping.

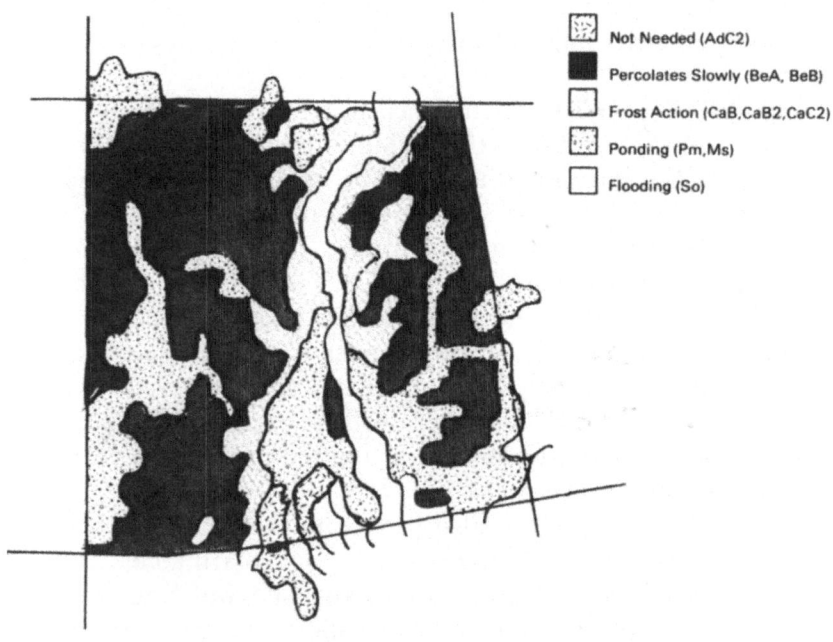

Figure 5-2. Drainage.

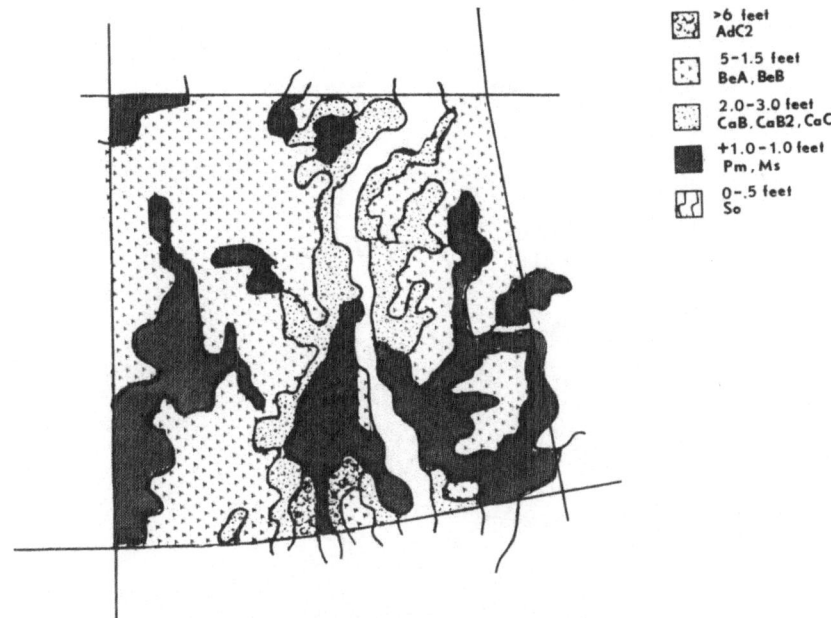

Figure 5-3. Depth to seasonal high water table.

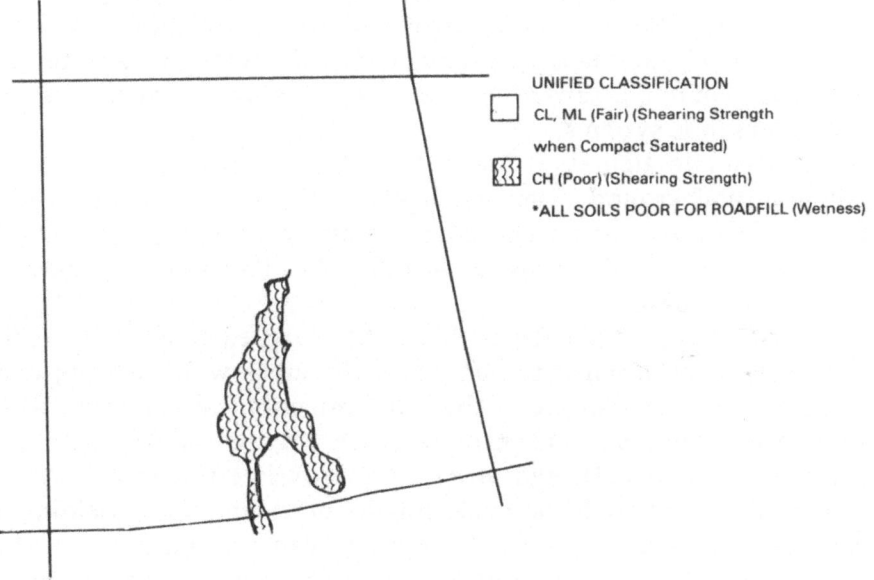

Figure 5-4. Bearing capacity.

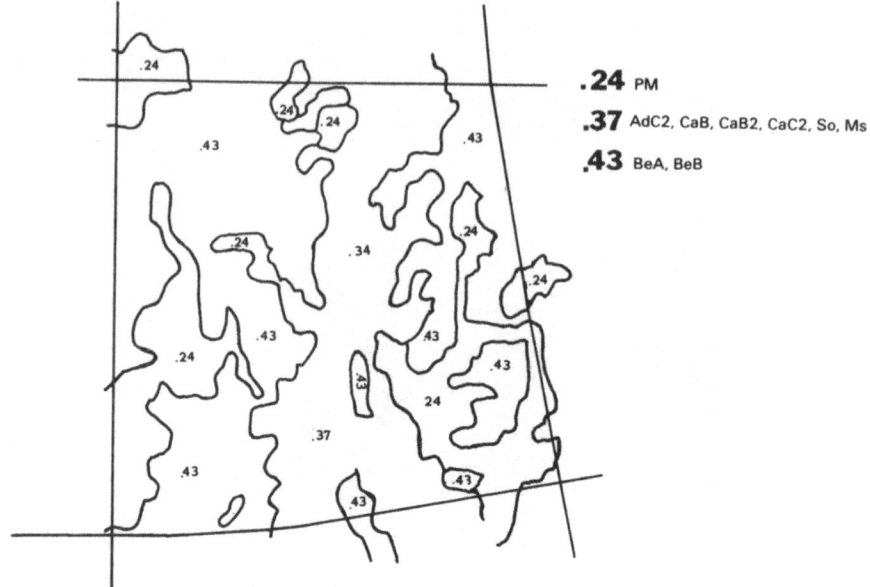

.24 PM

.37 AdC2, CaB, CaB2, CaC2, So, Ms

.43 BeA, BeB

Figure 5-5. Erosion potential "K" factor.

values of 0.37 and 0.43. A value of 0.49 is the maximum generally found. Finally, there are some areas with bedrock (Figure 5-6) very close to the surface causing problems with foundations, any future installation of public sewers and water, and the present installation of wells and septic systems.

The last two figures, 5-7 and 5-8, show respectively the flood potential and ground water availability. Both figures show conditions unfavorable for residential development: a high flood hazard and a rather low rate of water availability, 5 to 10 gallons per minute, over most of the site.

Following a method such as that of McHarg, each of these conditions should be given a rating. Thus, the areas with ponding and flooding could be assigned to the worst category (say, 7); those with slow percolation an intermediate score (5); and those with no drainage problems (1); and all given appropriate shades or colors. Similar decisions could be made for the other variables, yielding a composite map like Figure 5-9. Here, one can see that much of the site is rated poor for residential development for a combination of reasons: poor drainage, flooding, bedrock, and water availability.

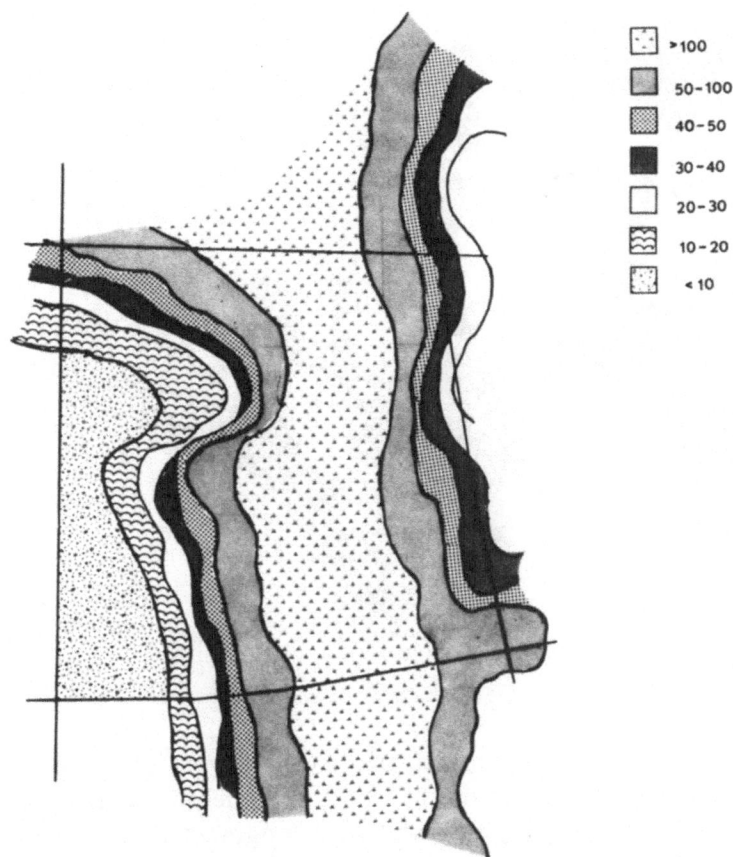

⬚	>100
▨	50-100
▦	40-50
■	30-40
☐	20-30
▨	10-20
⬚	<10

Figure 5-6. Depth of bedrock in feet.

Both the residents and environment would be adversly affected by the residential development of this area.

Critique. Fortunately, our example was rather straightforward. Conditions on this example site are so bad that the recommended land use decision becomes rather obvious: do not allow this kind of development. However, in this and in other less extreme examples, we can point to several major problems with capability analysis. Basically, we can criticize it for four major reasons:

1. *Subjectivity.* The selection of the variables of importance and the rating of those variables involve a very subjective process. In our

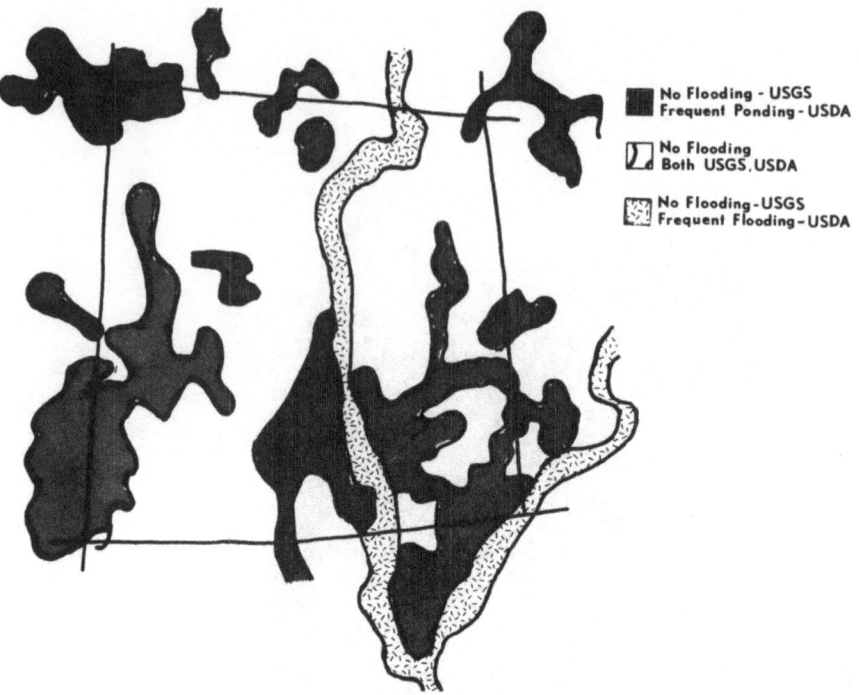

■ No Flooding - USGS
   Frequent Ponding - USDA

▨ No Flooding
   Both USGS, USDA

▧ No Flooding - USGS
   Frequent Flooding - USDA

Figure 5-7. Flood potential.

example, one must know a priori which types of soil conditions might produce problems. The extent of the problems is not measured quantitatively but rather is a judgment that problems will likely occur. The extent of the problem, the final ratings or rankings, could easily differ among various analysts causing the final recommendations to vary. Many professionals who use this technique fail to define adequately the rationale for choosing a particular ranking, making it difficult if not impossible for others to criticize the work constructively. The subjectivity problem becomes particularly difficult when it involves resources that could be used in different ways depending upon one's goals and objectives. For example, one can preserve agricultural land for continued agricultural use—the potential bias of an agriculturalist—or develop it for urban uses—the bias of a realtor. In addition, there are some natural resource problems that can be avoided altogether by investment in additional infrastructure or changes in design. Thus, septic systems may not be a problem if

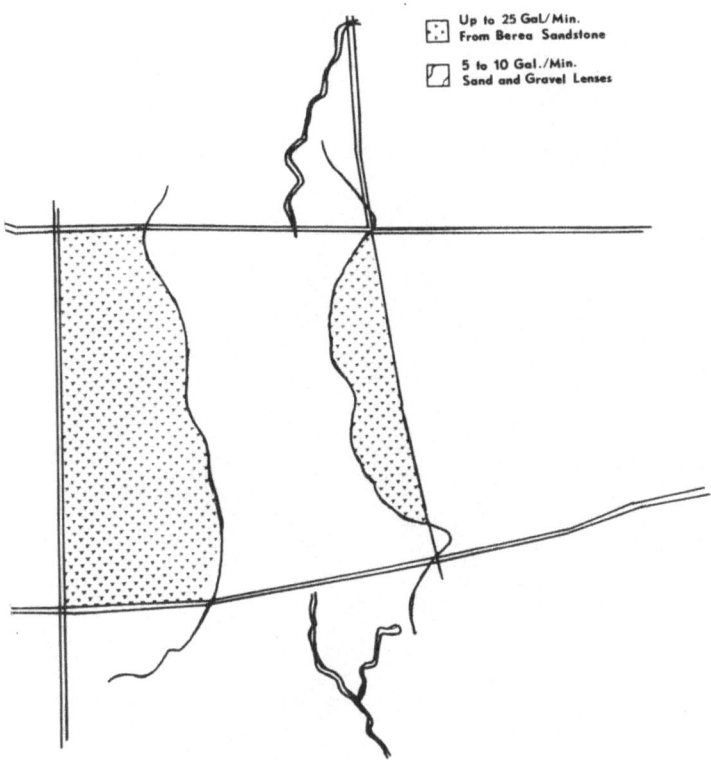

Up to 25 Gal/Min.
From Berea Sandstone

5 to 10 Gal./Min.
Sand and Gravel Lenses

Figure 5-8. Ground water availability.

one is willing to pay to extend existing sewer lines. Although no
process can be without subjectivity, capability analysis embeds so
many subjective decisions that it could become an unreliable tool.

2. *Problem of weighting.* The subjective decisions discussed above
are further exacerbated by the question of how important each
resource is relative to every other resource. Is it more important to
protect our land resources for agriculture or to preserve the same
land in its natural state for recreational uses? Are the impacts
associated with soil erosion more or less important than the
preservation of important historical and archeological sites? The
answers again vary with one's individual or group priorities.
However, the methods discussed above cannot easily be used in
any situation in which the weights are not equal across categories.
It is impossible to interpret maps visually that are twice as dark.

3. *The Blob.* No, we are not alluding to the "classic" movie with

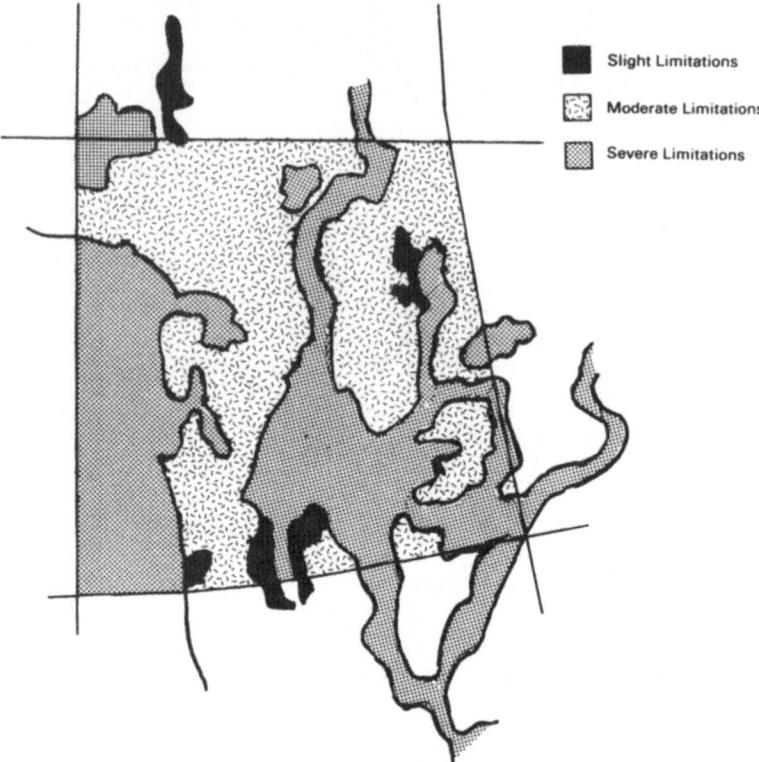

Figure 5-9. Composite map.

Steve McQueen but instead to the final product which one produces having to use more than five or six transparent overlays. Given this result, the overlay method easily falls short of the number of variables one might consider to be important.

4. *Ecological naiveté.* Somehow, those that continue to use overlay methods of capability analysis are always surprised when they "discover" geographic correlations among ecological variables. To those who have studied only a minimal amount of ecology, it should not be surprising that clay soils are associated with poor drainage characteristics or that coniferous forests are found in very well drained and not very fertile sandy soils. Yet, the always implied and often announced discovery of just such characteristics seems to accompany all capability analyses. This naiveté seems innocent enough on the surface, but it can greatly threaten the efficacy of any implementation based on such an analysis.

Geographic association does not always imply the same ecological mechanisms of cause and effect. The association between two variables could be coincidental or it could be caused by a third variable which is not being measured. The method in fact identifies no cause and effect relationships even though these are known to exist. Without such ecological understanding, the potential for unintended environmental impacts remains a very real possibility; Interesting examples of just such an occurrence are the Denver earthquakes (Healy et. al, 1971).

Capability analysis may in itself provide some data to avoid certain obvious impacts, but because of the simplicity of the method—which makes it popular and easy to use—it provides little or no possibility of advancing knowledge concerning the physical environment or of providing explicit information that can be used to manage the environment in more than a very superficial way.

AUTOMATIC GARBAGE OR HIGHTECH CAPABILITY ANALYSIS

With the advent of computer information systems aimed at analyzing large quantities of geographic data have come various attempts at automating the capability analysis. Rather than being transformed into maps by hand, the geographic data are converted to computer readable format, digitized, and the resulting information is processed using software packages that allow changes of scale, overlays, and output mapping with a line printer or plotter. The initial phase of such information system development can be characterized as advancing the state of the technology but not advancing the state of the art. What is meant by this, of course, is that automation of the analysis and mapping tasks does not change the *quality* of the results one iota although many of the uninitiated appear to think that it does.

Along with very simplistic information systems that repeat both the mistakes and good points of capability analysis, there have also come some innovations that have attempted to change the state of the art. The first of these is a relatively simple innovation in the scoring system associated with the analysis. In particular, many "modern" analysts have begun to use a linear weighted model to help describe the capability of a land unit. The linear weighted model has the following form:

$$C_{jk} = \sum_{i=1}^{n} W_{ik} \, X_{ijk}$$

where

C_{jk} = the final weighted index or score for the jth unit and kth utility value;

W_{ik} = weighting factor for the ith variable for kth utility value;

X_{ijk} = numerical value of ith variable for the jth unit, kth utility values;

n = number of variables used in the rating.

Use of this equation allows one to vary the weights associated with different variables and the scores associated with the values of particular variables. Thus, one can assign a utility value of 1 to well drained soils, 50 to those moderately drained, and 100 to those poorly drained, reflecting a major concern for such changes in environmental characteristics. Similarly, the weight of the drainage variable could be set to 2 while that for unique vegetation could be set to 1, a relative importance of only-half. By means of comparing final composite scores and mapping them, the problems of weighting and the blob become less difficult.

The problems of subjectivity and ecological naiveté are still not addressed by this approach. Indeed, the reality of the linear weighted model is that one must decide a priori what the weights will be in order to produce the "correct" results. What is correct is decided by the analyst using his or her experience with the situation as the bench-mark once again. Thus, if the final scores do not "reveal" the correct answer, the weights and utility values are experimentally manipulated until this answer is reached. Although most of the information systems we will review have not advanced beyond this stage (some actually have not reached this degree of sophistication), there have been some experimental attempts to employ other techniques. These are discussed in the next section.

SOME NEW EXPERIMENTS

Several researchers have experimented with quantitative techniques for the derivation of capability analysis information. The general approach to such analyses is borrowed from a set of techniques called *numerical taxonomy* (see Sneath and Sokal, 1973). Many of these techniques were originally used for the classification of biological species; they are based on the classification theories of Adanson. His approach to classification does not use a priori decisions about what differentiates one group of individuals from another. Rather, the strategy is to include enough information about the objects being classified in so far as technically possible. Further, the objects are classified according to their similarities rather than their differences. When this is done using a set of multivariate statistical techniques, one is performing a *numerical taxonomy*.

The general theory of numerical taxonomy applies to land capability analysis when this analysis represents a classification of the land units. The parallel becomes a little difficult in that the "individuals" being classified are geographically rather than genetically defined. The boundary that represents the difference between one "individual" and another thus becomes unclear and complicates the classification process. The choices for fixing such boundaries are basically two: (1) attempt to find the natural boundary between natural resource units, or (2) create an arbitrary unit of constant size. The first decision involves the creation of multiple, polygonal units of irregular shape and size such as soil units. The second involves the creation of a constant grid system and simply encoding the characteristics represented in that unit. The first decision creates relatively technical, data management problems while the second creates more theoretical questions as to what the data actually represent. The exact nature of these problems will be discussed with regard to the following examples.

An application of numerical taxonomy for land classification is provided by Gordon (1978). An area in Medford Township, New Jersey was, divided into 40 acre grid cells and data encoded on 42 variables. A list of these variables is given as Table 5-2. A two step procedure was used to analyze the relationships among the 42 variables and then to group a sample of the township into homogeneous land units.

Table 5-2. Variables for Medford Township, NJ

Soil Survey Variables	Natural Resource Inventory Variables
• available water capacity	depth to bedrock
• bearing capacity of soil	• fire hazard
• depth to water table	• flood hazard
• erodibility	• fog susceptibility
• frost-action potential	• microclimate
• liquid limit	• percentage area of high soil loss
• percent coarse fragments	(greater than 25 tons per acre per year)
• percent sand	• percentage area of low soil loss
• percent silt	(less than 5 tons per acre per year)
• permeability	• percentage area of medium soil loss
• plasticity index	(5–25 tons per acre per year)
shrink–swell potential	• percentage area of underlying
• slope degree	geological resource
• soil thickness	• percentage of scenic value
structure 1: platy	• percent bog
structure 2: prismatic	• percent cedar swamp
structure 3: columnar	• percent Cohansey formation
structure 4: blocky	• percent coniferous forest
• structure 5: granular–crumb	• percent deciduous forest
structure 6: structureless–massive	• percent floodplain
• structure 7: structureless–grains	• percent high humidity
type of fragments	percent in Medford
• woodland suitability	• percent mixed forest or successional forest
	• percent pine barrens
	• percent successional meadow
	• percent urban
	• percent Vincentown formation
	• percent wetland
	• summer-day comfort
	• summer-night comfort
	type of bedrock
	• wind exposure
	• winter-day comfort
	• winter-night comfort

Source: Gordon, 1978.

In the first step, a principal components factor analysis was used to delineate the interrelationships of the 42 variables, thereby explaining 70% of the variance via the use of 10 new factors. With this technique, "the factor loadings show the degree of relationship between each original variable and each new factor in an analogous manner to a correlation coefficient" (Gordon, 1978, p. 919). The factors, variables, and their loadings are shown as Table 5-3. Here, one can see

Table 5-3. Quantitative Relationships among Variables in Table 5-2

Factor	Name	Variables	Loading
1	soil characteristics, hydrology, pine barrens ecosystem	available water capacity	−0.678
		erodibility	−0.928
		fire hazard	0.749
		frost-action potential	−0.772
		liquid limit	−0.883
		percentage area of low soil loss	0.639
		percentage area of medium soil loss	−0.709
		coniferous forest	0.653
		percent sand	0.806
		percent silt	−0.911
		permeability	0.808
		plasticity index	−0.713
		structure 5: granular–crumb	−0.736
		structure 7: structureless grains	0.860
		woodland suitability	0.678
2	wetland humidity	fog susceptibility	0.908
		high humidity	0.912
		percent wetland	0.544
3	microclimate	low-stress microclimate	−0.678
		summer-day comfort	−0.693
		winter-night comfort	−0.699
4	slope	percent floodplain	−0.519
		percent mixed forest or successional forest	−0.708
		slope degree	0.609
5	deciduous forest geologic resource	percentage area of underlying geological resource	−0.791
		percent deciduous forest	−0.649
6	bearing capacity	bearing capacity of soil	0.738
		percent coarse fragments	0.615
7	scenic value	percentage area of low soil loss	−0.681
		percentage of scenic value	−0.815
8	bog	percent bog	0.743
9	cedar swamp	percent cedar swamp	−0.643
10	developed land	percent urban	0.856

Source: Gordon, 1978.

the strong quantitative relationships among environmental variables. Soil texture and structure are closely related to a number of vegetation types and to woodland suitability. Similarly, high humidity, fog susceptibility, and the percentage of wetland are highly correlated. Thus, the first step of the taxonomy tells us in more explicit, quantitative terms the nature and degree of relationships among the variables being used to evaluate the natural environment.

The second step of the taxonomy uses a classification program (Veldman, 1967) and the results of the first step to place the original grid cells into relatively homogeneous groups. This algorithm is a hierarchical classification system based on the measurement of the within groups' variance for the objects being classified. The model works iteratively, first joining those two cells which have the most in common and therefore produce the least impact on within groups variance. New members are added to groups one at a time until all groups merge into one heterogeneous group. For the Medford case, the group membership at the stage of 10 groups was arbitrarily selected as the stopping point for further evaluation. Each of the final 10 groups was examined to determine its central characteristics and the management techniques implied for that group. For example, the first group is all cells with high soil loss potential. Any development in these areas must proceed with caution to avoid problems with erosion, sedimentation, and related water quality impacts.

Similar experiments by other researchers have used other classification techniques. Rowe and Sheard use the discriminant analysis program available in the SPSS package to classify areas in the Northwest Territories of Canada (Nie et. al, 1970). They begin with a more traditional, a priori approach to classification claiming this is more correct theoretically, and then test their classification with numerical taxonomy. (Rowe and Sheard, 1981)

Other experiments along this vein have been undertaken. It is safe to say that at present the use of numerical techniques for classification is more an art than a science. Thus, the practitioner must fall back on the more traditional approaches to capability analysis except under special circumstances. Below, we review some of the major issues associated with the computerization of this process and then give some recent examples.

ISSUES AND PROBLEMS WITH AUTOMATED LAND CAPABILITY CLASSIFICATION

Polygonal Overlays

The overlay technique for land capability analysis has been discussed above. One might recall that the method is very subjective, does not allow for the different weighting of variables, and fails to indicate the interrelationships among the variables being analyzed. Yet, the method remains in wide use because of its simplicity and the benefits of undertaking such an analysis in the absence of any other. In fact, to be fair, we must indicate that an overlay analysis does yield a good amount of useful information about the environment and the potential impacts of development upon it. When we look at these same problems and benefits in the context of computerization of the technique, many of the same issues arise but several, additional, more technical problems come forward as well. Thus, this section attempts to review those issues before we go on to look at examples of such computer information systems.

The first problem which arises when transferring geographic information to the computer is one of data representation. An excellent review of many of these issues including a description of some major systems is given by Dueker (1979). The first such issue is whether to represent the geographic information in polygonal or grid encoded formats. Polygonal representations are advantageous in that the boundaries of natural entities such as soils, vegetation, topography, and so on, are irregular polygons. People are more accustomed to viewing geographic information in this form in that it conforms more exactly with the data and with the traditional maps of the overlay method. Encoding of a polygon does create some technical problems at the boundaries where spaces or slivers emerge due to the imperfections of the digitizing process. This flaw has required the creation of some rather sophisticated software to prevent such occurrences. More problematic for environmental analysis is the calculation of overlay datasets from several sets of variables. Extremely sophisticated and not widely available routines are needed to keep track of the location, nature, and amount of

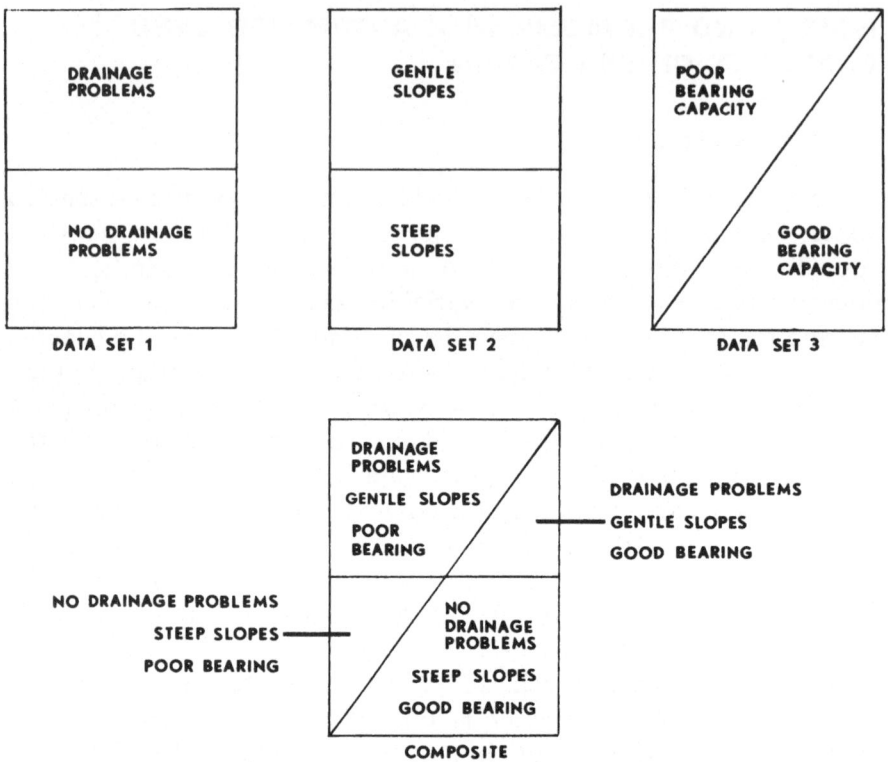

Figure 5-10. Problems with polygonal overlays.

overlap among categories of variables. Figure 5-10, using the overlay of three rather simple datasets, illustrates the difficulty of this task.

Grid Encoding

Mostly because of the analytical problems associated with polygonal overlays, but also for other reasons, grid encoding has become more popular for natural resource information systems. Other reasons include ease of encoding, with or without a digitizer, elimination of the need for a plotter unit for map output, and potential for use of the data in various statistical analyses. With grid encoding, natural boundaries are approximated by regular, rectangular grids thus slightly altering the actual boundaries. The inaccuracy thus created depends on the grid size and input data scale and is discussed further on.

Overlays created through grid encoding become more straightforward, tabulating processes where one counts the number of cells in which various combinations of variables occur. This is illustrated in Figure 5-11. Output can easily be represented on a line printer and does not require a plotter. Figure 5-12 is such a map. Although encoding is faster with a digitizer, one can simply create an overlay grid base map, place this on other maps, and encode values by hand onto a keypunch or other data entry system. This process reduces the total cost of the information system.

Finally, the grid cells can become the "individuals" input to a numerical analysis such as the Medford Township example discussed above. Such an analysis is not possible with polygons unless they are converted to some type of grid format.

Problems with Grid Encoding.   A number of ancillary issues arise with the encoding process. Perhaps the most important of these is the question of information *loss*. This is most problematic for grid encoding systems. Such losses arise because of the encoding scale as it relates to the scale of the input data. If, for example, the input resolution of a soils map is approximately one acre but a 40 acre grid cell is used in the information system, information must be reduced in the encoding process. First, the generalizations along the boundary areas will automatically cut off sections of the natural soils polygons represented on the base map. Second, the system must make a decision regarding which soil should be assigned to the grid cell. Decision rules often prescribe that the entity comprising the majority of the cell will be assigned to that cell. One can easily imagine a situation where a soil will only take up 35% of a cell with the remainder being made up of 5% to 15% of a number of other soils. Assigning the total cell to one soil causes all additional information to be lost.

An alternative decision rule could assign an estimated cell percentage to several categories of soils. Here, the amount of encoding which must be done increases proportionally, with many of the encoded values being zero. Also, the overlay becomes problematic since one ends up with many internal combinations of cell traits which cannot be mapped at the scale in which the system operates. The result of this information loss can be critical, causing a major resource problem to be overlooked.

1. Select Maps to be Overlaid

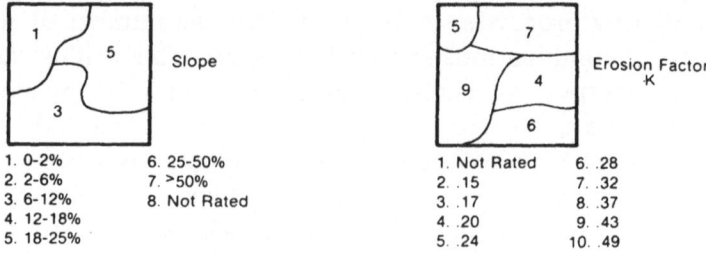

1. 0-2%	6. 25-50%
2. 2-6%	7. >50%
3. 6-12%	8. Not Rated
4. 12-18%	
5. 18-25%	

1. Not Rated	6. .28
2. .15	7. .32
3. .17	8. .37
4. .20	9. .43
5. .24	10. .49

2. Determine the categories for the final map

Erosion Potential

1. No erosion	5. 3-4 tons
2. < 1 ton per acre per year	6. >4 tons
3. 1-2 tons	7. Not Rated
4. 2-3 tons	

3. Determine the combinations of slope and erosion to produce the desired categories.

Slope Code	→	matched with	→	Erosion Factor Code	→	will result in	→	Erosion Potential Code
1				1				7
1				2				1
1				3				1
1				4				1
1				5				2
1				6				2
1				7				2
1				8				2
1				9				2
1				10				2
↓		Continue matching codes		↓				↓
7				10				6
8				-1*				7

* Indicates that any erosion code with slope code 8 will be assigned a code 7, "Not rated"

4.
The combinations are fed into the map program, and the computer compares the slope file and the erosion factor "K" file, and produces the new file with the appropriate codes.

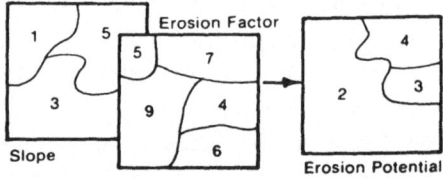

Figure 5-11. Overlay based on grid encoding (OCAP). (Source: G. Gordon, 1978.)

The best examples of the impact of such information loss can be illustrated by focusing on a proposed housing development in northern New Jersey. This highly sloped area posed many development problems that were easily discernible by viewing a standard, U.S. Geological Survey $7\frac{1}{2}$ minute quadrangle map with a contour

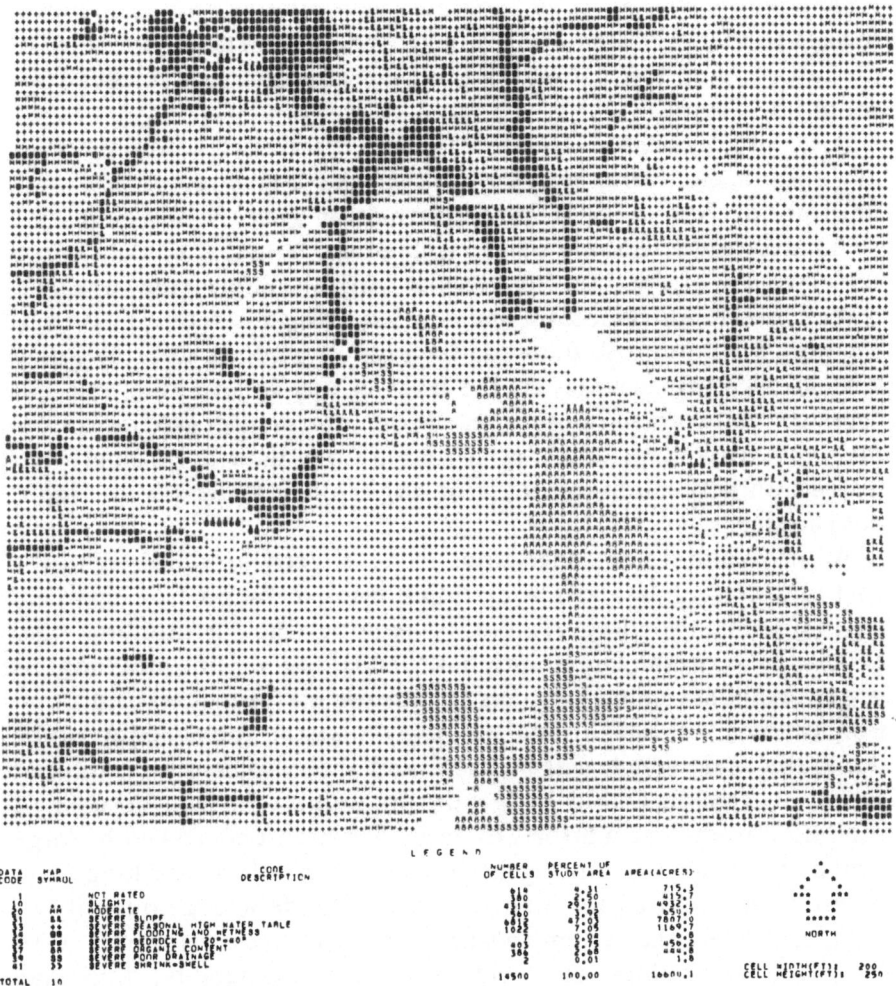

Figure 5-12. Limitations for homesites with basements. Soil Survey, Harrisville, Township, Medina County, Ohio. (Source: Gordon and Gordon, 1981. Reprinted with permission.)

interval of 20 feet. Contour lines bunched closely together indicate very high slopes. The developer, probably realizing that this did not look good for his presentation, changed the vertical scale of the map to a 100 foot contour interval. Slopes looked much better since one line now represented what five had done before. Unfortunately, the developer also used this map for planning purposes. Accordingly, a location was selected for the sewage treatment facility, set between two of the 100 foot contours. A check in the field revealed that the real location was on a 90 foot hill, necessitating the pumping of sewage to that elevation if the project were built. The implications of the information loss caused by the generalization to a smaller scale should be obvious.

A second problem related to encoding choice stems from the scale of input data. Here, the problem is the converse of that with information loss, namely false information *gain*. If one chooses a small area grid size based on the scale of the largest scale data input, the accuracy of the data for items at larger scale becomes exaggerated. Thus, if one chooses a one acre grid which is determined to be accurate for input soils data and then inputs geologic information at a scale allowing accuracy at the level of one square mile, one has "created" data which do not exist. Most information systems do not take the trouble to label input and output data accordingly but instead blithely pump out maps at *any* scale regardless of the input data source.

All of the above considerations relate directly or indirectly to the cost of encoding data and analyzing it using a computer. Regardless of the scale, encoding remains a tedious, time consuming, and expensive process. Until automatic encoding of line drawings is widely available (some products are already on the market), information systems will have to make compromises on the encoding to be done, the scale which will be used, and the output to be produced simply to remain within budgetary limitations. In fact, because of such limitations, most efforts to date have been undertaken by large urban, regional, or state agencies which can afford the long term investments required to build a system, encode a large quantity of data, and bring the system to some useful purposes. However, with improved technology and wider software availability, this may change in the future.

Analytical Weaknesses of Land Information Services

Regardless of the encoding method used, the analytical capabilities of most land information systems remain lacking. If one uses an overlay approach, the same judgmental and subjectivity problems which arise with non-computerized approaches continue to occur. In addition, many systems are even more limited in scope, only allowing simple changes in scale or input/output routines. What is most surprising is that almost none of these systems have attempted to link the encoded data with any of the more analytical environmental models. For example, one could easily link soils data with data on hydrology and use the information as input to projections of the stormwater runoff impacts of future development (see Fromuth, 1978, for an excellent example). Similarly, land use data in the information would allow better forecasts of both water and air emissions. These outputs could then be linked to one of the air or water quality models.

Although some attempts have been made in these directions, the systems that incorporate various modelling techniques tend to be narrowly focused on one type of analysis. For example, the PEMSO system of W. E. Gates and Associates (OEPA, 1980) allows only a generalized downslope tracking of sewage waste load allocations for the siting and sizing of new sewer lines but does not allow any capability analysis to be undertaken. Similarly, the OCAP system of the Ohio Department of Natural Resources (Gaybrielle Gordon, 1978) allows extended data manipulation for capability analysis but currently is not connected to any environmental models.

Even if the more sophisticated statistical techniques of numerical taxonomy are used to augment standard capability analysis, the state of the art leaves much to be desired. The choice of statistical techniques alone can be as subjective as the choice of weights in more traditional analyses. The complexity of the methods requires a very high level of expertise in order to avoid misinterpretations. Though the techniques appear to hold future promise, they are not developed well enough for present, practical use.

Finally, one must recognize a number of practical problems associated with system implementation. None of the available information systems do exactly what everyone would like. Thus, in addition to the resources for encoding and analyzing information, any potential new user of such a system must have the ability to reprogram the system. Computer programmers experienced in this area are not very easy to find or inexpensive to hire. Given that one can produce the needed output, problems of servicing the system users also arise. Few individuals can decide a priori which final maps or analyses they would like to see. With a batch oriented or expert-user oriented computing environment, new requests are answered with great time delays, defeating the major usefulness of the information system—quick information turn-around.

MODEL REVIEWS

Two land information systems will be compared and contrasted as examples of the capabilities of such systems. These are certainly not the only such systems nor do they represent all possible combinations of system configurations. However, they are important because they have been widely applied and tested and because they have stood the test of time—both are still used. Both systems are also available under licensing agreements to other users. The systems we are referring to are OCAP (for Ohio Capability Analysis Program) and IMGRID (for Information Management on a Grid Cell). Each is reviewed below.

Ohio Capability Analysis Program

OCAP was developed by the Ohio Department of Natural Resources in 1973 (G. Gordon, 1978). It uses a modified grid encoding system called a *raster system.* Here, rather than storing repetitive data on equal size grids, OCAP digitizes along constant scan lines called rasters. Only the locations of changes in information, the boundaries between categories, are encoded at their location relative to the Northwest corner of the map being digitized. This is illustrated by Figure 5-13. The smallest geographic unit generally used in OCAP studies is approximately 1.1 acre of resolution. The system also allows for the storage of an attribute matrix relative to any mapped data with multiple attributes. The most prominent example is soils data where one boundary system is related to a large number of important physical and chemical soils properties. These properties can be stored in matrix form rather than requiring separate digital input.

The typical data used in an OCAP study is shown in Table 5-4. From this table one can see that a wide array of physical and cultural data are encoded for possible further analysis. The studies generally exclude a number of hydrologic, atmospheric, and effluent data that would be required for other types of environmental modelling. This does not preclude the system handling such information, which is not included simply because no one has requested it or knows how to use it if it were included.

The OCAP program is operated only in batch mode on a main-

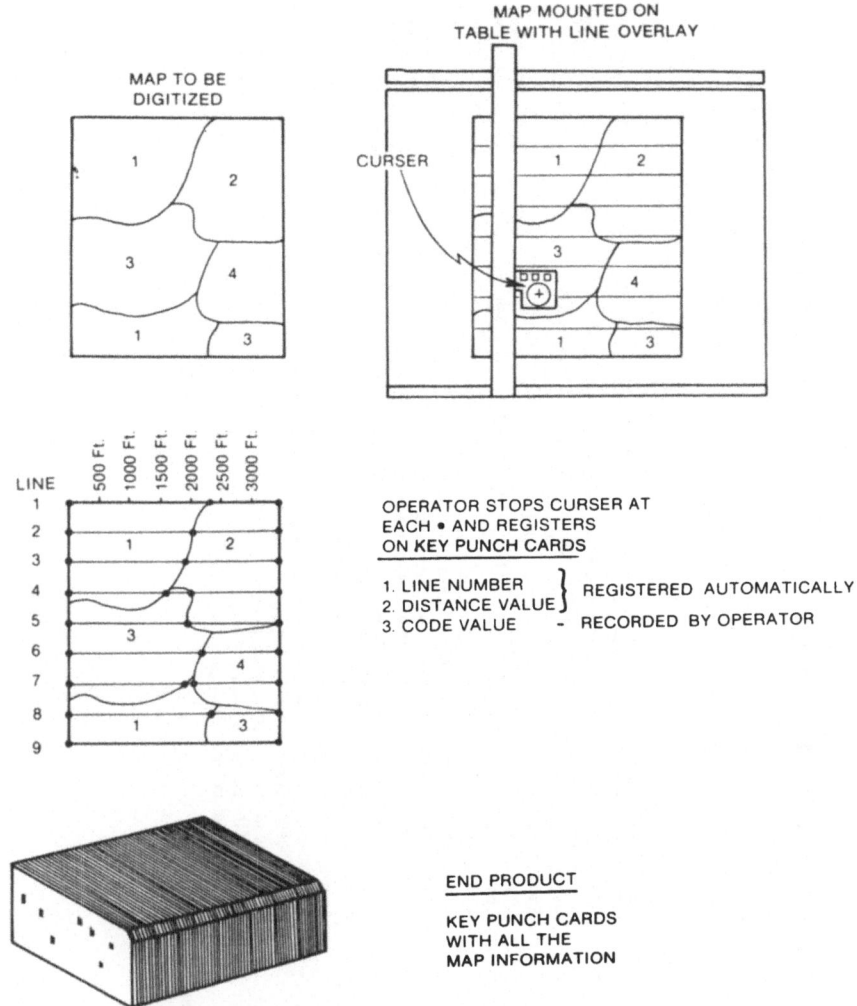

Figure 5-13. Digitizing process. (Source: G. Gordon, 1978.)

frame computer. It is written in PL-1, which is mainly available on IBM systems. This may be problematic for transfer to other systems. The program itself is divided into eight major programs with various functions. Data are passed between programs via disk files. The programs are listed in Table 5-5 along with their major functions. The *edit* program is used for various housekeeping chores including corrections to digitized maps and additions and deletions from data

Table 5-4. OCAP Information Included in a Typical Study

Map Name	Description	Source
Soil Map Units	Combinations of soil type and percent of slope mapped in the detailed Soil Survey. Soil evaluations are for the earth's surface to 5 feet below.	USDA Soil Survey
Soil-Related Maps		
Susceptibility to Flooding	Relative likelihood of flooding based on soil evidence of previous flooding. (may be improved by using flood hazard maps)	USDA Soil Survey
Drainage Class	An indication of the rapidity and extend of removal of water from soil.	USDA Soil Survey
Depth to Seasonal Water Table	An indication of the shallowest water table conditions resulting from an impervious layer or saturation expected to occur at some time each year.	USDA Soil Survey
Agricultural Capability	Shows the relative usefulness of a soil for agriculture.	USDA Soil Survey
Potential Frost Action	Likely danger from excessive wetting, heaving, and loss of strength due to frost.	USDA Soil Survey
Depth to Bedrock	Depth of soil overlying bedrock.	USDA Soil Survey
Erosion Factor "K"	A measure of potential erosion loss from a soil type.	USDA Soil Survey
Erosion Factor "T"	A measure of the acceptable soil loss from a soil type.	USDA Soil Survey
Shrink-swell Potential	A measure of the volume change of a soil in response to a change in soil moisture.	USDA Soil Survey
Unified Class	A classification system for soil which evaluates bearing strength for heavy loads.	USDA Soil Survey
AASHO Class	American Association of State Highway Officials Classification system — an evaluation of the soil for highway construction.	USDA Soil Survey
Corrosion Potential Steel	An evaluation of the relative rate that uncoated steel corrodes in soil.	USDA Soil Survey
Corrosion Potential Concrete	An evaluation of the relative rate that concrete corrodes in soil.	USDA Soil Survey
Permeability Rate	A measure of the ease with which soil transmits water downward (inches per hour).	USDA Soil Survey
Texture Class	A classification system based on soil particle size.	USDA Soil Survey
Available Water Capacity	A measure of the soil water content available for plant use.	USDA Soil Survey
Slope	Measure of the percent of incline of land.	USDA Soil Survey
Limitations for: Agriculture Homesites (3 stories or less) Sewage Effluent Disposal Sewage Lagoons Lawns, Landscaping Streets & Parking Lots Athletic fields Park & Play Areas Campsites - Tents Campsites - Trailers Sanitary Landfills - Trench type	Soil Conservation Service ratings of the soil limitations for each land use.	USDA Soil Survey

Table 5-5. OCAP Computer Program Overview

Program	Function	How It Can Be Used
Edit	Correct and update map files	To update information; to add information to base data files; to change information in the base data files.
Map	Map and analyze base files	To produce maps of the information with titles, symbols, and descriptions to suit the analyst. Analysis using a linear weighted model; overlays.
Matrix	Add auxiliary information to data files	To extend the use of some base data files, particularly soils.
Merge	Merge files into larger units	To put together several small files into a larger or more complete one.
Premod	Combine single variable map files	To combine several files into a single file in preparation for mapping.
Boundary	Extract information for particular boundaries.	To map information for any boundary such as political, watershed, or census tract boundaries. Can also be used with boundaries created by search.
Scale	Rescale any base data files	To make county or other large maps a more manageable size.
Search	Create new boundaries within a certain distance of a point, line, or area	To look in the vicinity of a land use of concern to determine the proximity of problem areas.
Ogre	Produce tables of statistics for one, two, or three base files	To produce tabular listings of information in the base files when statistics, rather than maps, are needed.

After Gaybrielle Gordon, 1978, p. 42.

files. The *map* program is one of the most interesting. Not only does it allow for the creation of line printer maps of any variable in the database but it also allows for the overlay of up to 30 variables at one time, the tracking of combined categories, and the use of a linear weighted model which creates a combined score of each land parcel across all overlayed attributes. Next, the *matrix* program allows for the storage of auxiliary data such as the soils data mentioned above. *Merge* is another data handling program which allows for the combination of small files with the same variables into a larger file such as USGS quad based files being merged into a county file. The *premod* program would be used as a prelude to a map or overlay routine where variables from different files are being compared. The *boundary* program allows for a very flexible reformatting of the geographic information to any political or natural boundaries that are digitized by the user. As the name implies, the *scale* program allows for the rescaling of the data-base to any size. *Search* produces a new database that can subsequently be mapped, that shows the area a fixed distance from a line or within a circle of user supplied radius. This allows for the analysis of resources within the vicinity of a major new facility that might have an impact on the resources. Finally, the *ogre* program, as its name implies, gobbles up data and spits out tabular listings of the statistical contents of a file. Surprisingly enough, many programs currently in use do not allow for the compilation of such simple statistics. Using ogre, one can find out how many acres of a particular unit exist in the study area, how many acres of a combination of categories exist in an area (previously obtained by an overlay), and so on.

OCAP offers an overview of a number of desirable features of land information systems but also involves a number of problems. The programs themselves are technically sound but potentially inaccessible to other users because they are PL-1 based and require a very large computer. The technology of map review and production is also quite antiquated, requiring the exclusive use of batch processing and the production of maps only with a line printer. This is very problematic in that such maps do not allow for printing various common reference points such as roads, streams, points of interest, and so forth. Thus, one must produce mylar overlays with such features on them to aid the map user. Newer systems should be designed for more transportability as well as to allow one to take advantage of newer, terminal based processing for editing and for map production on graphics printers or plotters.

A second set of problems arises from the lack of linkage of the OCAP program to any, more analytical models of environmental quality. As was mentioned in Chapter 3, some research has been done on ways of linking OCAP to various stormwater models (Fromuth, 1978) but no effective action to implement such a linkage has been taken. The effort required to encode detailed land use and soils information is only partly effective if one cannot take full advantage of a range of environmental applications with the data. More effort needs to be expended in this direction.

Finally, OCAP suffers from all of the subjectivity and analytical problems associated with the overlay and land capability analysis techniques. Users must define, by trial and error, the weights and utility values associated with each variable to arrive at a set of overall scores which make intuitive sense. The analysis remains more an art than a science.

Overall, OCAP has come farther than many other information systems in that it provides a full array of data manipulation programs not available elsewhere. As such, it represents one of the better programs of this type and can be used to set some *minimum* standards for future developments in this field.

The IMGRID Program

The Information Management on a Grid Cell System (IMGRID) was originally developed by researchers in the Graduate School of Design at Harvard University and is marketed by the Harvard Laboratory for Computer Graphics (Harvard Laboratory for Computer Graphics and Spatial Analysis, 1978). Other versions have been developed by the Tennessee Valley Authority, Office of Natural Resources, and by ERDAS Incorporated of Atlanta, Georgia. This latter version is written for a microcomputer system called the ERDAS system (ERDAS, undated). IMGRID is batch process oriented and produces line printer maps which have the same appearance as OCAP maps. The ERDAS version is interactive rather than batch oriented. This is advantageous, but its advantages are offset in that none of the statistical routines available in the batch version are present in the interactive version.

Data for IMGRID are basically hand encoded into a constant grid system. Depending on the grid cell size chosen, it thus suffers from information loss problems. The user must incorporate his or her own decision rules for assigning a particular value to a grid cell (for example, whether 50% of the grid is occupied by a particular variable). In some versions of the model, only a limited number of codes are available for each variable (9 for some purposes in the TVA model or 99 for others; 15 in the ERDAS version for all purposes). This is a problematic limitation for such data as soil type since over 100 may occur in a particular county. The user must then prescreen data in several ways and make many encoding judgements that can bias later analysis results and also produce large information losses. (Such decisions are not necessary in the OCAP model. Still, some information loss does occur because of the scale at which data were encoded.) Encoding must also be done at constant scale. This implies that all information must be transferred to a common base map before encoding takes place—also a negative aspect of the model, requiring a large amount of additional labor.

IMGRID is keyword controlled. Each keyword indicates that a particular calculation is to be performed on a variable or set of variables. There are 51 such keyboards in the TVA version of the program. It is beyond the scope of this example to review the purpose of each of these keywords. Thus, we have selected those that are involved in the analysis tasks most important to land capability analysis. This listing is followed by an overall critique of the program. The important keywords for capability analysis follow:

Overlay. Unlike its counterpart in the OCAP map program, this overlay routine does not produce composite maps which summarize across a number of characteristics using a linear weighted model. Instead, this routine, with a 19 variable maximum, examines the ordinal scale codes (which must have values from 0 to 9 *only*) and reassigns the highest code present across all variables to each of the grid cells being analyzed. Thus, one would have to record previously all variables to the nine categories in such a way that the maximum code indicates the presence of a critical variable.

Recode. This allows for the recoding of any variable into a new set of nominal, ordinal, or ratio scale values. This must be used to reassign allowable ranges of values to variables undergoing analysis with limited subroutines such as the overlay routine.

Matrix. On first blush this appears to be a more sophisticated recoding program wherein one examines the values of two variables and creates a new, third, variable based on the nature of the first two. In fact, this is done but with a rather strange twist. Rather than using a simple reassignment process, matrix assigns a new value as a function of the formula

matrix entry value + ((row value − 1)

$$\times \text{ number of matrix columns}) + \text{column value}$$

The reasons for this are completely unclear in the documentation. The effect is that the analyst must take a significant amount of time, better spent on other things, to figure out the numerical result of particular combinations.

Summary and **Table.**   These keywords result in data tabulations for selected variables, giving the user a breakdown of the number of cells in each category and the cross-tabulation of the values of two variables respectively. This is very useful information that is not, unfortunately, available in the ERDAS version of the program. Also, the programs are limited in their scope with comparisons available for only two variables at a time based on the table function, and one at a time for the summary. Thus, cross-tabulation of many variables that may be examined in a land capability analysis is extremely cumbersome, if not impossible.

Add, Index, Mult, Sub.   These functions can be used to help create a pseudo-linear weighted model. The user can add, subtract, or multiply a particular value by a scalar or weight and place the result in the dataset as a new variable. Unfortunately, all but the index routine can be used only with one or two variables at a time. The index routine can handle up to 20 variables but requires the recoding or prior rescaling of all input variables as a separate step, making the process very cumbersome.

Search.   There are actually four search routines built into the IMGRID program. Each searches in the proximity of one or more cells. The search pattern differs with the keyword used and can be circular, from a point, a band along a line, in a diamond or square pattern from a point, and in a user defined functional distance based on some variable such as slope.

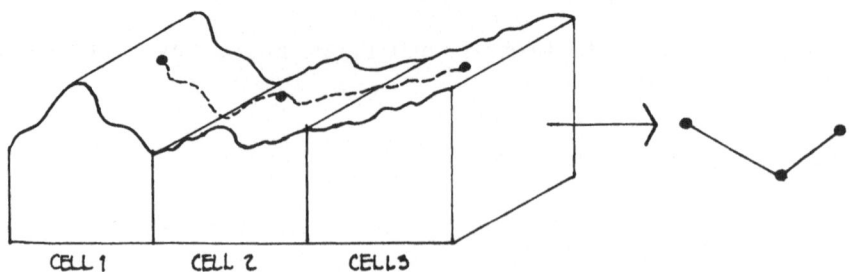

Figure 5-14. Profile routine illustration (IMGRID).

Profile. This routine produces a vertical profile of slope based on an input matrix of elevation values. The profile does not produce a map but rather a side view transect across a map. This is illustrated in Figure 5-14.

The preceding review of the characteristics of the IMGRID system should make its good and bad features apparent. The user must exercise a good deal of judgement in making coding decisions and must perform many calculations by hand in order to ensure that the results "make sense." Not only is there significant information loss, but the amount of work necessary to undertake straightforward computer based analyses may be so large as to make it easier to perform the analysis by hand. The system does allow for relatively low cost input/output mapping operations, requiring neither a digitizer nor a plotter unit. However, this equipment is becoming quite inexpensive and offers the benefits of more automatic input and better output quality respectively.

REFERENCES

Adanson, M. (1757). *Histoire naturelle du Senegal.* Cited in Sneath, Peter H. A. and Robert R. Sokal (1973). *Numerical Taxonomy.* New York: W. H. Freeman.

Dueker, Kenneth J. (1979). "Land Resource Information Systems: A Review of Fifteen Years of Experience," *Geo-Processing,* Vol. 1, p. 105–128.

Earth Resources Data Analysis Systems (undated). *User's Manual.* ERDAS, 999 McMillan Street, Atlanta, GA 20218.

Fromuth, Rick (1978). *Selection of Stormwater Models for Application to The Ohio Capability Analysis Program.* Columbus, OH: Ohio Department of Natural Resources.

Gordon, Gaybrielle (1978). *User's Guide to the Ohio Capability Analysis Program.* Columbus, OH: Ohio Department of Natural Resources.

Gordon, S. I. (1978). "Performing land-capability evaluation by use of numerical taxonomy: land use and environmental decisionmaking made hard?" *Environment and Planning A,* Vol. 10, p. 915–921.

Gordon, Steven I. and Gaybrielle E. Gordon (1981). "The Accuracy of Soil Survey Information for Urban Land Use Planning," *APA Journal,* July 1981, p. 301–312.

Harvard Laboratory for Computer Graphics and Spatial Analysis (1978). *LAB-LOG.* Cambridge, MA: Harvard University Press.

Harvard University, Department of Landscape Architecture (1967). *Three Approaches to Environmental Resource Analysis.* Washington, DC: The Conservation Foundation.

Healy, J. W., W. W. Rubey, D. T. Griggs, and C. B. Raleigh (1971). "The Denver Earthquakes," in *Man's Impact on Environment,* ed. by Thomas R. Detwyler. New York: McGraw-Hill Book Co.

Hills, G. Angus (1966). "The Classification and Evaluation of Land for Multiple Uses," *Forestry Chronicle,* June 1966, p. 1–25.

Holmes, David D. and Rebecca L. Jolly (1980). *IMGRID,* version 3.5. Norris, TN: Tennessee Valley Authority.

McHarg, Ian (1969). *Design with Nature.* New York: Natural History Press.

Lewis, Philip H. (1964). "Quality Corridors for Wisconsin," *Landscape Architecture Quarterly,* January, p. 100–107.

Nie, Norman H. et al. (1975). *Statistical Package for the Social Sciences, SPSS.* New York: McGraw-Hill Book Co.

Ohio Environmental Protection Agency (1980). *The PEMSO Nonpoint Source Screening Report No. 2.* Columbus, OH: Office of the Planning Coordinator, OEPA.

Rowe, J. Stan and John W. Sheard (1981). "Ecological Land Classification: A Survey Approach," *Environmental Management,* Vol. 5, No. 5, p. 451–464.

Sneath, Peter H. A. and Robert R. Sokal (1973). *Numerical Taxonomy.* New York: W. H. Freeman.

Veldman, Donald J. (1967). *Fortran Programming for the Behavioral Sciences.* New York: Holt, Rinehart and Winston.

6
Hazardous Wastes

INTRODUCTION

One of the newest areas of environmental control and concern is that of hazardous wastes. This concern has come from the major health risks associated with exposure to many of the orgnaic and inorganic chemicals produced to serve our industrial, agricultural, and other domestic needs and from the sheer volume of solid wastes, including hazardous wastes, that are generated in the U.S. In 1976, municipal solid wastes, amounted to about 130 million metric tons, "enough to fill the New Orleans Superdome from floor to ceiling, twice a day, weekends and holidays included." (USEPA, May 1978). Estimates of industrial waste generation are 344 million metric tons per year of which 10–15% is probably hazardous waste. This concern has been magnified by the relatively recent "discovery" and publicity associated with abandoned hazardous waste facilities near water supplies, residential developments, schools, and so on.

There are more than 100,000 surface impoundments containing industrial wastes in the U.S. Thus, the nature of the disposal problems of hazardous wastes involves both the cleanup of abandoned waste dumps that pose a threat to public health and the proper siting and construction of new hazardous waste treatment facilities. As we will see, each of these tasks must account for a number of complex and poorly understood chemical processes and environmental conditions in order to arrive at a recommended course of action. Models which attempt to address these questions are thus relatively new, untested, and incomplete. Yet, without such models the analyst cannot hope to arrive at intuitive decisions that will protect the environment and public health.

There are several pieces of legislation at the Federal level which have promulgated much of the work in the hazardous waste disposal area. The Resource Conservation and Recovery Act (Public Law 94-580, October 21, 1976), or RCRA, governs the disposal of all solid wastes in the U.S. Solid waste is defined very broadly as all solid, liquid, semisolid, and contained gaseous material resulting from essentially all human activities. RCRA has resulted in a number of policies associated with the siting of new waste disposal facilities and governing the disposal of all newly generated wastes. The Toxic Substances Control Act of 1976 (Public Law 94-469) provides for the control of the production of hazardous materials through agency and industry testing of the potential risks associated with both "old" and "new" chemicals as defined by the act and with authority to ban the production of chemicals found to cause "unreasonable risk of injury to health or the environment." (Environmental Law Reporter, ELR 41341). The Comprehensive Environmental Response, Compensation, and Liablity Act of 1980 (otherwise known as the "Superfund" legislation) provides for the establishment of a fund to clean up releases of hazardous substances from spills and from abandoned waste disposal sites.

Each of these pieces of legislation provides for a complex set of regulatory powers and has resulted in a number of implementing regulations at the Federal and state levels whose discussion could comprise a book in and of themselves. The reader is referred to other sources for such discussions (see *Environmental Law Reporter* and related discussion papers and regulations cited in the end-of-chapter references. For our purposes it is enough to note that the combination of these and other pieces of Federal and State environmental legislation has effectively resulted in "cradle to the grave" control of hazardous substances. Thus, the estimation of the potential impacts associated with waste cleanup and waste disposal options becomes critical to the rational implementation of the regulations.

INTERTWINED ENTRY PATHS FOR HAZARDOUS WASTES

The modelling of the impacts of hazardous wastes is obviously of great importance. Unfortunately, the nature of the environmental interactions involved in this process makes this modelling very difficult. This is illustrated by Figure 6-1. Beginning either with the movement of old, buried chemical wastes or with the introduction of new chemicals, hazardous materials enter the environment through a number of pathways. With regard to new chemicals, a review process is introduced as a requirement of the Toxic Substances Control Act

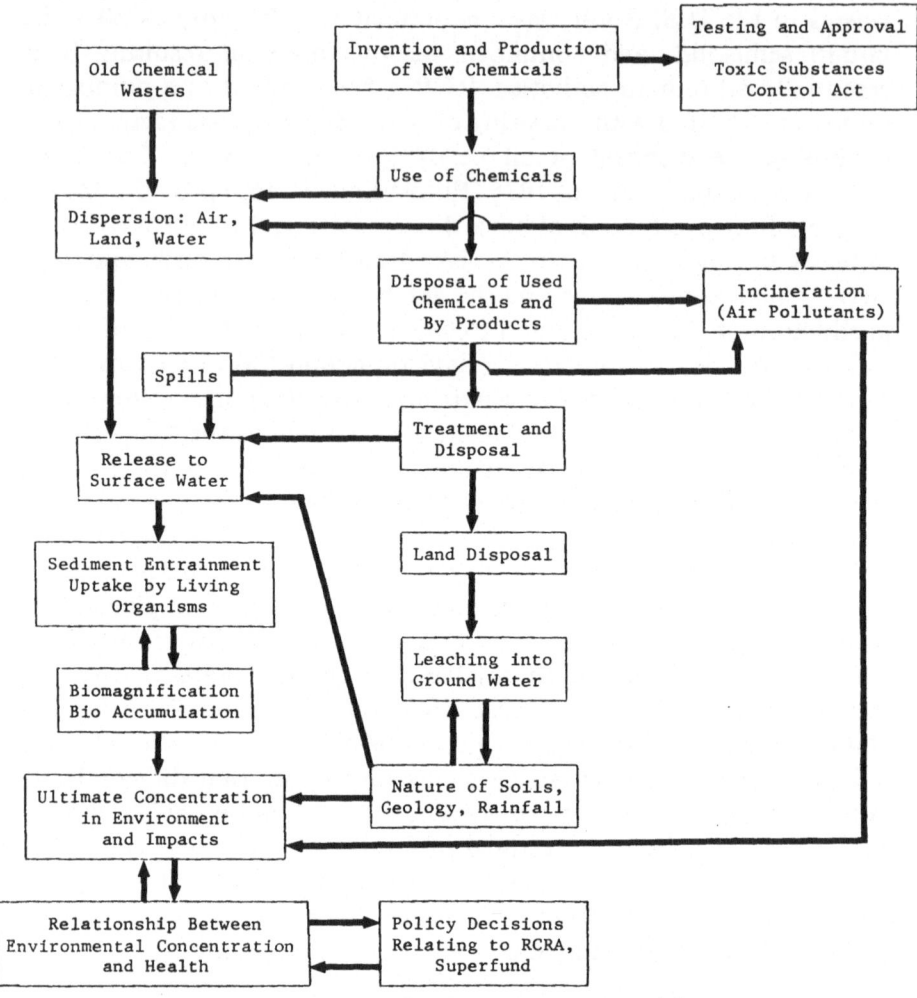

Figure 6-1. Complexities of hazardous waste modeling.

to prevent chemicals found to be unacceptably hazardous from being manufactured for commercial use. The use of those that are acceptable, in manufacturing, agriculture, or domestic activities will introduce some residual chemicals into the environment, a further undesirable effect.

Any unused chemicals, residues, or by-products will have to be disposed of in a manner acceptable under the Resource Conservation and Recovery Act. This could be by incineration, or other treatment with land disposal. Thus, the chemicals could enter the air, water, or solid waste streams. Similarly, older chemical wastes may begin to enter the environment as they escape from open dumps and burial sites. In either case, the first set of uncertainties develops because the type, rate, mechanisms, and timing of the environmental releases are unknown and, perhaps, unknowable.

Once a release has occurred, the propagation path that the pollutants will follow also varies greatly. In the case of land disposal, most experts seem to now agree that an absolutely secure landfill is impossible. Once release takes place, the variability of the environment complicates the calculation of the rate of movement of the pollution plume. The nature of the soils and geologic formations will vary widely from place to place; that is, these materials are not homogeneous. The greater the heterogeneity of these formations, the more variable will be the transmission of the pollutants. This effect is further complicated by the potential uptake of some of the chemicals through the process of adsorption, the rate of which vary with the materials involved and the concentration of the pollutants. Finally, the rate of the overall movement will be affected by the amount of leaching, which in turn is related to the amount of rainfall over time.

The surface water fate of hazardous pollutants is also complex. Pollutants can be diluted, absorbed onto sediments which settle to the bottom, then stirred up with the sediments and released into the water; or as water moves to and from the ground water system, they can interact with that system as contributors or they can receive further pollutants from that system. If the pollutants remain in the water, they may combine with other chemicals or be taken up by the aquatic vegetation and plants. As a result of such uptake, the concentration may accumulate in selected organisms or actually increase in concentration because of their solubility in the fatty tissues of certain plants and animals (biomagnification or bioaccumulation). Thus, the ultimate concentration of these pollutants in the ambient environment is very difficult to estimate. Given such concentrations, it is also extremely difficult to isolate the health impacts of these substances at less than extreme levels. In the face of these uncertainties, the policy decisions which one might make relating to RCRA and the Superfund often involve more art than science.

The remainder of this chapter reviews these many concepts then analyzes several models which have been used to investigate the impacts of hazardous wastes.

THE FATE OF HAZARDOUS WASTES

Given the difficulties noted above with tracing the movement of hazardous chemicals in the environment, how should we approach the modelling of these systems? Some are convinced that no models are better then *any* computer models when it comes to these matters (Baski, 1979). These individuals feel that professional judgement, or intuition, is much better at yielding a reliable result. The fate to which we are referring, a suitable term to go with intuition, is that of chemicals and includes their movement in the natural environment. This fate can be subdivided into that in surface waters and in ground water systems.

In surface waters, there are thought to be six distinct fates for chemicals. These include photolysis, hydrolysis, volatilization, sorption, bioaccumulation, and biotransformation/biodegradation (USEPA, 1979). The transport or pathway of each chemical may be different because of its particular characteristics, reactivity with environmental chemicals, and uptake by organisms in the natural environment. Table 6-1 summarizes each of these processes and the chemicals and processes that most affect them.

In the ground water system, two major processes impact the concentration of chemical wastes. These are (1) the dilution of the chemicals caused by their diffusion and mixing within the underground aquifer and (2) the adsorption or sorption of the chemicals by earth materials. This latter process is similar to the one that occurs in the aquatic environment. It is generally estimated using the *Freundlich isotherm equation*:

$$C_s = K_p C_w^{1/n}$$

where

C_s = the concentration of the chemical in the particulate phase;
C_w = the concentration of the chemical in the water phase;
K_p = a partition coefficient for sorption; $1/n$ exponential factor.

Table 6-1 Chemical Transport Processes in Surface Water

Process	Definition	Influences on Process
Volatilization	Movement of chemical vapor to the atmosphere from the water surface	High vapor pressure Low solubility
Sorption	Attachment of chemical to bottom and suspended sediments	More hydrophobic chemicals
Photolysis	Photochemical transformation from direct or indirect chemical excitation in the presence of sunlight	No generalizations available
Oxidation	Reaction with oxygen due to direct or indirect photolysis or reaction with oxidizing chemicals.	No generalizations available
Hydrolysis	Combination with the the hydroxyl ion in water	Related to acidity
Bioacumulation	Concentration of substance in certain species, especially in fatty tissues of fish and in bone marrow	Related to fat and lipid solubility and inorganic chemicals partitioned into other body organs
Biotransformation and Biodegradation	Conversion to other chemical forms by microbial population	Related to microbial population and chemical concentrations

For particular chemicals and particular types of sediments, the coefficients can be measured in laboratory experiments. The nature of the sorption is influenced by the fraction of organic carbon in the sediment as well as the amount of clay and related inorganic compounds with high sorptive capacities.

The movement of water through an underground aquifer is governed by the equations of *Darcy's law*. One form of this equation is

$$V = -K \frac{dh}{dL}$$

where

V = the velocity of groundwater flow;
K = the permeability coefficient;
dh = the difference in head between the ends of the aquifer;
dL = the difference in length over which the movement occurs.

This law is based on the principle that groundwater will flow through tortuous paths formed by interconnected interstices of the aquifer particles. The flow will be slow and therefore will be laminar. The exact definition of the permeability coefficient has been open to some question since it involves the characteristics of the porous medium and that of the fluid (Ward, 1967, p. 273). Thus, Theis has proposed the use of the term *transmissibility,* which "quantitatively describes the ability of the aquifer to transmit water." (As cited in Ward)

In order to use this law for the modelling of groundwater flows, one must make a number of further assumptions concerning the nature of the flow. For flow in two dimensions, the initial models relating to these phenomena have assumed that the flow is in an artesian aquifer (a confined aquifer under pressure) which is non-homogeneous and isotropic (general flow characteristics are the same in all directions from a point). Under these conditons, the partial differential equation for the flow is

$$\frac{\partial}{\partial x}\left(\frac{T\partial h}{\partial x}\right) + \frac{\partial}{\partial y}\left(\frac{T\partial h}{\partial y}\right) = \frac{S\partial h}{\partial t} + Q$$

where

>   T = aquifer transmissivity;
>   h = head;
>   t = time;
>   S = aquifer storage coefficient;
>   Q = net groundwater withdrawal rate per unit area;
> x, y = rectangular coordinates.

The solution to the equation can be approximated using either finite difference or finite element approximations (see Faust and Mercer, 1980). When one of these methods is employed, the aquifer is divided into a number of grid cells with nodes representing the locations for which the system of finite difference equations will be solved. The nodes can either be mesh-centered, that is on the grid intersections, or block centered, in the middle of each cell. Figure 6-2 illustrates a block centered grid system. When such a system is used, the model is solved for the water movement in the system given a set of boundary conditions, changes in recharge, and changes in pumpage rates throughout the system.

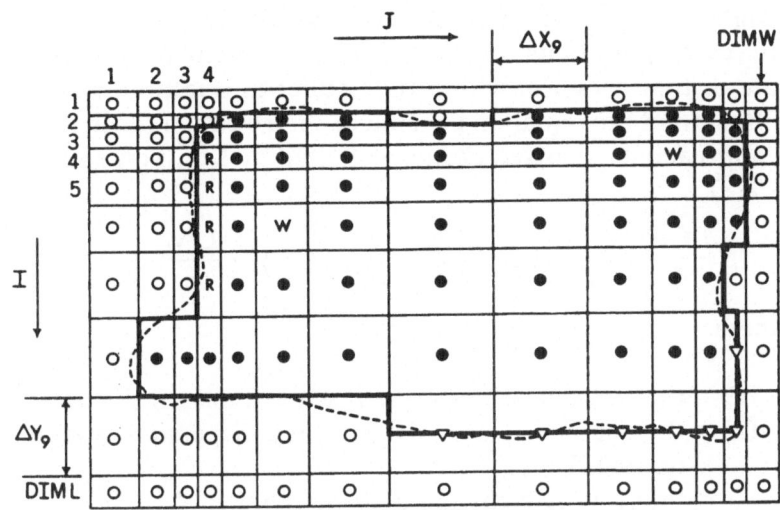

EXPLANATION

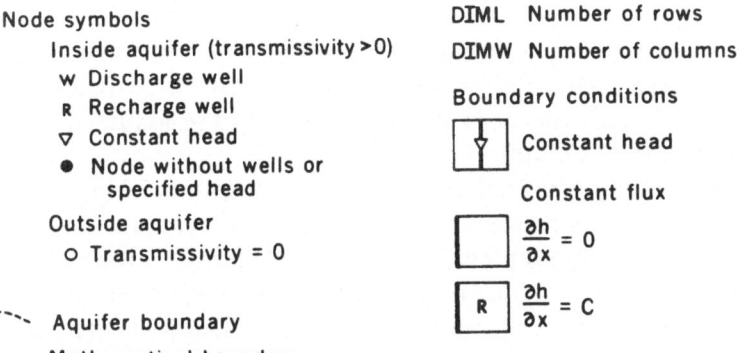

Node symbols

  Inside aquifer (transmissivity >0)

  w Discharge well

  R Recharge well

  ∇ Constant head

  ● Node without wells or
    specified head

  Outside aquifer

  o Transmissivity = 0

---- Aquifer boundary

—— Mathematical boundary

DIML Number of rows

DIMW Number of columns

Boundary conditions

  ∇ Constant head

  Constant flux

  $\frac{\partial h}{\partial x} = 0$

  R $\frac{\partial h}{\partial x} = C$

Figure 6-2. Finite-difference model for aquifer simulation. (Source: Trescott, Pinder and Larsen, 1976, p. 31.)

The boundary conditions of the aquifer have a major impact on the model results. In general, models of groundwater flow tend to delineate the boundaries of the aquifer far enough away from the area of concern to allow one to ignore the effects of the boundary itself.

The flow of the groundwater is only one part of the model that simulates the fate of hazardous materials. In conjunction with this flow, one must calculate the diffusion of any solutes within the aquifer. Toward this end, one must ideally know the location of the source of pollutants and the rate of pollutant release, the sorption rate of that chemical within the aquifer in question, and the rate of diffusion of the remaining chemical. An equation for the transport of a nonreactive dissolved chemical is cited by Konikow and Bredehoeft (1978):

$$\frac{\partial(Cb)}{\partial t} = \frac{\partial}{\partial x_i}\left(bD_{ij}\frac{\partial C}{\partial x_j}\right) - \frac{\partial}{\partial x_i}(bCV_i) - \frac{C'W}{\epsilon} \qquad i, j = 1,2$$

where

C = the concentration of the dissolved chemical species, M/L^3;
D_{ij} = the coefficient of hydrodynamic dispersion (a second order tensor), L^2/T;
b = the saturated thickness of the aquifer, L;
C' = the concentration of the dissolved chemical in a source or sink fluid, M/L^3

Some models of solute dispersion simply do not account for any chemical reactivity. Others represent sorption in terms of a retardation factor. This factor is approximated by the assumed adsorption isotherm for the material in question.

Conceptually, the fate of hazardous substances in the aquatic and groundwater environments do attempt to account for the many mechanisms of movement and chemical change. However, in most instances, the data required to estimate the parameters imbedded in the equations above become problematic. Data on the application rates or environmental concentrations of hazardous substances are sparse. Obtaining such information is both difficult and expensive, requiring complex chemical tests, and in the case of groundwater, difficult sampling problems. Data on actual water withdrawals from aquifers must also be estimated while information on the geologic formations must be deduced indirectly from old well records or expensive sampling programs. Thus, even if the models briefly described above are accurate representations of reality, estimation of the required parameters for model operation is very difficult and subject to errors.

Because of the nature of the models discussed in this chapter, the examples that are used are somewhat different than was the case in previous chapters. The examples are based on rather simple situations in order to demonstrate the nature of both the input and output associated with each model. The policy questions that were posed in previous examples are not used here because of the difficulties in using these models for that purpose. This does not mean that no such policy analyses are possible. However, the model results illustrated in this chapter cannot be viewed in the same predictive framework that is the case with several of the previous reviewed types of models.

MODEL REVIEWS

The Hazard Ranking System

Under the provisions of the Comprehensive Environmental Response and Liability Act, the USEPA and the States must determine which abandoned hazardous waste sites and new, emergency situations, such as spills, involve the greatest risk for the environment (used broadly here to mean the ecosystem and human health) as an aid to making decisions concerning allocations of the "Superfund." Partly because of the lack of explicit data on these sites (loading rates, groundwater hydrology, and so on) and partly to provide a means of responding to a very critical situation as rapidly as possible, the Agency has promulgated a technical ranking system to assess the risk arising from hazardous waste sites.

This system is not computerized. Rather it is an organized way of obtaining and analyzing limited data on the sites and arriving at a score representing the relative hazard of each site with regard to groundwater, surface water, and air route contaminations of the environment. It may be useful to review this system here as an indication of the state-of-the-art in hazardous waste cleanup determination.

The Hazard Ranking System (HRS) is part of the regulations promulgated under CERCLA (Federal Register, Vol. 47, No. 137 [July 16, 1982]). The system is "a procedure for ranking facilities in terms of the potential threat they pose by describing:

- The manner in which the hazardous substances are contained,
- The route by which they would be released,
- The characteristics and amount of the harmful substances, and
- The likely targets."

HRS results in a composite score for each site as follows:

$$S_M = \frac{1}{1.73} \sqrt{S_{gw}^2 + S_{sw}^2 + S_a^2}$$

where

S_{gw} = ground water route score;
S_{sw} = surface water route score;
S_a = air route score.

The number is scaled to a 100 point scale. The nature of the data used to derive each rating is summarized in Table 6-2. Here, one can see that the system attempts to be comprehensive in nature by considering route characteristics, containment, waste characteristics, and potential targets (i.e. locations impacted by the hazardous material). Each item is considered relative to a specific indicator for which the analyst must use field data on the site to enter a rating table and obtain a score. For example, the route characteristics for groundwater are based on the depth to the aquifer measured as the vertical distance between the lowest point of hazardous waste and the highest seasonal level of the aquifers saturated zone. Table 6-3 shows the values that are assigned relative to these distances.

Similarly, values are assigned for net precipitation (greater rainfall bringing added hazard), permeability of the geologic materials (more permeability bringing a higher score), and the physical state of the hazardous material (solids are less problematic than liquids or gas). If the movement of the wastes may be contained due to natural or artificial conditions, or, more important, will provide no containment, additional points are added.

The nature of the wastes is also evaluated in terms of persistence, toxicity, and quantity. Tables 6-4 to 6-6 illustrate these ratings. Finally, the distance to the nearest well and the population served (at risk) determine a score for the target. The scores are then tabulated and multiplied by a set of constants to rescale them to a consistent level to arrive at the final score for the groundwater route. This is illustrated in Figure 6-3.

A similar approach is followed for both surface water and air and then combined using the formula given above to yield a combined score from 0 to 100. This number can be used as a *relative* indication

Table 6-2. Comprehensive List to Rating Factors

Hazard Mode	Factor Category	Factors		
		Ground Water Route	Surface Water Route	Air Route
Migration	Route Characteristics	• Depth to aquifer of concern • Net precipitation • Permeability of unsaturated zone • Physical state	• Facility slope and intervening terrain • One year 24-hour rainfall • Distance to nearest surface water • Physical state	
	Containment	• Containment	• Containment	
	Waste Characteristics	• Toxicity/Persistence • Hazardous waste quantity	• Toxicity/Persistence • Hazardous waste quantity	• Reactivity/Incompatibility • Toxicity
	Targets	• Ground water use • Distance to nearest well/population served	• Surface water use • Distance to sensitive environment • Populaton served/distance to water intake downstream	• Land use • Population within 4-mile radius radius • Distance to sensitive environment

Table 6-2. Comprehensive List to Rating Factors (cont.)

Hazard Mode	Factor Category	Factors		
		Ground Water Route	Surface Water Route	Air Route
Fire and Explosion	Containment	• Containment		
	Waste Characteristics	• Direct evidence • Ignitability • Reactivity • Incompatibility • Hazardous waste quantity		
	Targets	• Distance to nearest population • Distance to nearest building • Distance to nearest sensitive environment • Land use • Population within 2-mile radius • Number of buildings within 2-mile radius		
Direct Contact	Observed incident	• Observed incident		
	Accessibility	• Accessibility of hazardous substances		
	Containment	• Containment		
	Toxicity	• Toxicity		
	Targets	• Population within 1 mile radius • Distance to critical habitat		

Table 6-3. Score for Depth to Aquifer

Distance (feet)	Assigned Value
>150	0
76 to 150	1
21 to 75	2
0 to 20	3

Source: *Federal Register*, July 16, 1982, p. 31224.

Table 6-4. Containment Value for Ground Water Route

Assign containment a value of 0 if: (1) all the hazardous substances at the facility are underlain by an essentially non permeable surface (natural or artificial) and adequate leachate collection systems and diversion systems are present; or (2) there is no ground water in the vicinity. The value "0" does not indicate no risk. Rather, it indicates a significantly lower relative risk when compared with more serious sites on a national level. Otherwise, evaluate the containment for each of the different means of storage or disposal at the facility, using the following guidance.

	Assigned Value

A. Surface Impoundment

Sound run-on diversion structure, essentially nonpermeable liner (natural or artificial) compatible with the waste, and adequate leachate collection system	0
Essentially non permeable compatible liner with no leachate collection system; or inadequate freeboard	1
Potentially unsound run-on diversion structure; or moderately permeable compatible liner	2
Unsound run-on diversion structure; no liner; or incompatible liner	3

B. Containers

Containers sealed and in sound condition, adequate liner, and adequate leachate collection system	0
Containers sealed and in sound condition, no liner or moderately permeable liner	1
Containers leaking, moderately permeable liner	2
Containers leaking and no liner or incompatible liner	3

Table 6-4. Containment Value for Ground Water Route (cont.)

	Assigned Value
C. Piles	
Piles uncovered and waste stablized; or piles covered, waste unstabilized, and essentially non permeable liner	0
Piles uncovered, waste unstabilized, moderately permeable liner, and leachate collection system	1
Piles uncovered, waste unstabilized, moderately permeable liner, and no leachate collection system	2
Piles uncovered, waste unstabilized, and no liner	3
D. Landfill	
Essentially non permeable liner, liner compatible with waste, and adequate leachate collection system	0
Essentially non permeable compatible liner, no leachate collection system, and landfill surface precludes ponding	1
Moderately permeable, compatible liner, and landfill surface precludes ponding	2
No liner or incompatible liner; moderately permeable compatible liner; landfill surface encourages ponding; no run-on control ...	3

Source: *Federal Register*, July 16, 1982, p. 31229.

Table 6-5. Waste Characteristics Values for Some Common Chemicals

Chemical/ Compound	Toxicity[1]	Persistence[2]	Ignitability[3]	Reactivity[3]
Acetaldehyde	3	0	3	2
Acetic Acid	3	0	2	1
Acetone	2	0	3	0
Aldrin	3	3	1	0
Ammonia, Anhydrous	3	0	1	0
Aniline	3	1	2	0
Benzene	3	1	3	0
Carbon Tetrachloride	3	3	0	0
Chlordane	3	3	*0	*0
Chlorobenzene	2	2	3	0
Chloroform	3	3	0	0
Cresol-O	3	1	2	0

Table 6-5. Waste Characteristics Values for Some Common Chemicals (cont.)

Chemical/ Compound	Toxicity[1]	Persistence[2]	Ignitability[3]	Reactivity[3]
Cresol-M&P	3	1	1	0
Cyclohexane	2	2	3	0
Endrin	3	3	1	0
Ethyl Benzene	2	1	3	0
Formaldehyde	3	0	2	0
Formic Acid	3	0	2	0
Hydrochloric Acid	3	0	0	0
Isopropyl Ether	3	1	3	1
Lindane	3	3	1	0
Methane	1	1	3	0
Methyl Ethyl Ketone	2	0	3	0
Methyl Parathion in Xylene Solution	3	Δ0	3	2
Naphthalene	2	1	2	0
Nitric Acid	3	0	0	0
Parathion	3	Δ0	1	2
PCB	3	3	Δ0	Δ0
Petroleum, Kerosene (Fuel Oil No. 1)	3	1	2	0
Phenol	3	1	2	0
Sulfuric Acid	3	0	0	2
Toluene	2	1	3	0
Trichlorobenzene	2	3	1	0
∝-Trichloroethane	2	2	1	0
Xylene	2	1	3	0

[1]Sax, N. I., Dangerous Properties of Industrial Materials, Van Nostrand Rheinhold Co., New York, 4th ed., 1975. The highest rating listed under each chemical is used.

[2]JRB Associates, Inc., Methodology for Rating the Hazard Potential of Waste Disposal Sites, May 5, 1980.

[3]National Fire Protection Association, National Fire Codes, Vol 13, No. 49, 1977.

*Professional judgment based on information contained in the U.S. Coast Guard CHRIS Hazardous Chemical Data, 1978.

ΔProfessional judgment based on existing literature.

Toxicity	Assigned Value
Sax level 0 or NFPA level 0	0
Sax level 1 or NFPA level 1	1
Sax level 2 or NFPA level 2	2
Sax level 3 or NFPA level 3 or 4	3

Table 6-5. Waste Characteristics Values for Some Common Chemicals (cont.)

Tons in Cubic Yards	Number of Drums	Assigned Value
0	0	0
1–10	1–40	1
11–62	41–250	2
63–125	251–500	3
126–250	501–1,000	4
251–625	1,001–2,500	5
626–1,250	2,501–5,000	6
1,251–2,500	5,001–10,000	7
>2,500	>10,000	8

Value for Toxicity	Value for Persistence			
	0	1	2	3
0	0	0	0	0
1	3	6	9	12
2	6	9	12	15
3	9	12	15	18

Persistence of each hazardous substance is evaluated on its biodegradability as follows:

Substance	Assigned Value
Easily biodegradable compounds	0
Straight chain hydrocarbons	1
Substituted and other ring compounds	2
Metals, polycyclic compounds and halogenated hydrocarbons	3

Source: *Federal Register,* July 16, 1982, p. 31229–31230.

of the severity of the risks associated with a particular site. However, it does not actually measure these risks.

The problems associated with this approach should be obvious. Each factor is given an arbitrary equal weight. One can imagine a number of actual circumstances in which the environmental risk from one factor will far outweigh that from the others. For example, an extremely toxic chemical probably should be cleaned up regard-

Table 6-6. Persistence (Biodegradability) of Some Organic Compounds*

Value = 3   Highly Persisent Compounds

aldrin
benzopyrene
benzothiazole
benzothiophene
benzyl butyl phythalate
bromochlorobenzene
bromoform butanal
bromophenyl phyntyl ether
chlordane
chlorohydroxy benzephenone
bis-chloroisoprophyl ether
m-chloronitrobenzene
DDE
DDT
dibromobenzene
dibutyl phthalate
1,4-dichlorobenzene
dichlorodiffuoroethane
dieldrin
diethyl phthalate
di(2-ethylhexyl)phthalate
dihexyl phthalate
di-isobutyl phthalate
dimethyl phthalate
4,6-dinitro-2-aminophenol
dipropyl-phythalate
endrin

heptachlor
heptachlor epoxide
1,2,3,4,5,7,7-heptachloronorbornene
hexachlorobenzene
hexachloro- 1,3-butadiene
hexachlorocyclohexane
hexachloroethane
methyl benzothiazole
pentachlorobiphenyl
pentachlorophenol
1,1,3,3-tetrachloroacetone
tetrachlorophenyl
thiomethylbenzothiazole
trichlorobenzene
trichlorobiphenyl
trichlorofluromethane
2,4,6-trichlorophenol
triphenyl phosphate
bromodichloromethane
bromoform
carbon tetrachloride
chloroform
chloromochloromethane
dibromodichloroethane
tetrachloroethane
1,1,2-trichloroethane

Value = 2   Persistent Compounds

acenaphthylene
atrazine
(diethyl) atrazine
barbital
borneol
bromobenzene
camphor
chlorobenzene
1,2-bis-chloroethoxy ethane
b-chloroethyl methyl ether
chloromethyl ether
chloromethyl ethyl ether
3-chloropyridine
di-t-butyl-p-benzoquinone
dichloroethyl ether
dihyrocarvone
dimethyl sulfoxide
2,6-dinitrotoluene

cis-2-ethyl-4-methyl-1,3-dioxolane
trans-2-ethyl-4-methyl-1,3-dioxolane
guaiacol
2-hydroxyadiponitrile
isophorone
indene
isoborneol
isoprophenyl-r-isopropyl benzene
2-methoxy biphenyl
methyl biphenyl
methyl chloride
methylindene
methylene chloride
nitroanisole
nitrobenzene
1,1,2-trichloroethylene
trimethyl-trioxo-hexahydrotriazine
iosmere

Table 6-6. Persistence (Biodegradability) of Some Organic Compounds (cont.)

Value = 1   Somewhat Persistent Compounds

acetylene dichloride	limonene
behenic acid, methyl ester	methane
benzene	methyl ester of lignoceric acid
benzene sulfonic acid	2-methyl-5-ethyl-pyridine
butyl benzene	methyl naphthalene
butyl bromide	methyl palmitate
e-caprolactam	methyl phenyl carbinol
carbon-disulfide	methyl stearate
o-cresol	naphthalene
decane	nonane
1,2-dichloroethane	octane
1,2-dimethoxy benzene	octyl chloride
1,3-dimethyl naphthalene	pentane
1,4-dimethyl phenol	phenyl benzoate
dioctyl adipate	phthalic anhydride
n-oecane	propylbenzene
ethyl benzene	1-terpineol
2-ethyl-n-hexane	toluene
o-ethyltoluene	vinyl benzene
isodecane	xylene
isoprophyl benzene	

Value = 0   Nonpersistent Compounds

acetaldehyde	methyl benzoate
acetic acid	3-methyl butanol
acetone	methyl ethyl ketone
acetophenone	2-methylpropanol
benzoic acid	octadecane
di-isobutyl carbinol	pentadecane
docosane	pentanol
eicosane	propanol
ethanol	propylamine
ethylamine	tetradecane
hexadecane	n-tridecane
methanol	n-undecane

*JRB Associates, Inc., *Methodology for Rating the Hazrd Potential for Waste Disposal Sites,* May 5, 1980.

Source: *Federal Register,* July 16, 1982, pp. 31229–31230.

less of how far it is from an aquifer or how close to a population simply because leaving it in place will increase the risk with time. The HRS does not project the actual environmental damages or health risks associated with a spill or an abandoned hazardous waste site

Ground Water Route Work Sheet					
Rating Factor	Assigned Value (Circle One)	Multi-plier	Score	Max. Score	Ref. (Section)
① Observed Release	0　　　　　45	1		45	3.1
If observed release is given a score of 45, proceed to line ④. If observed release is given a score of 0, proceed to line ②.					
② Route Characteristics					3.2
Depth to Aquifer of Concern	0　1　2　3	2		6	
Net Precipitation	0　1　2　3	1		3	
Permeability of the Unsaturated Zone	0　1　2　3	1		3	
Physical State	0　1　2　3	1		3	
Total Route Characteristics Score				15	
③ Containment	0　1　2　3	1		3	3.3
④ Waste Characteristics					3.4
Toxicity/Persistence	0　3　6　9　12　15　18	1		18	
Hazardous Waste Quantity	0　1　2　3　4　5　6　7　8	1		8	
Total Waste Characteristics Score				26	
⑤ Targets					3.5
Ground Water Use	0　1　2　3	3		9	
Distance to Nearest Well/Population Served	} 0　4　6　8　10 12　16　18　20 24　30　32　35　40	1		40	
Total Targets Score				49	
⑥ If line ① is 45, multiply ① x ④ x ⑤ If line ① is 0, multiply ② x ③ x ④ x ⑤				57,330	
⑦ Divide line ⑥ by 57,330 and multiply by 100　　$S_{gw}=$					

Figure 6-3. Ground water route work sheet. (Source: *Federal Register*, July 16, 1982, p. 31225)

and thus a low score may give a false impression of low risk, which could very well not be the case.

A qualitative rating system such as this may in fact be required since the data and models are not available to assess a large number of sites in more detail. In other words, the state-of-the-art is not advanced enough to allow for more specific risk assessment. Only after longer experience with the nature of the movement of hazardous wastes, the damages caused by them, and the mechanisms by which they might better be controlled will more explicit modelling of risks be possible. Unfortunately, there is a large population potentially at risk in the interim.

USGS Solute Transport and Dispersion Model

A number of computer programs have been developed in an effort to make more explicit projections of the chemical concentrations of solutes in groundwater. One of the most accessible and better documented models comes from the U.S. Geological Survey (Konikow and Bredehoeft, 1978). The model is based on the general two dimensional flow and transport equations presented above. As such, it makes a number of assumptions, listed by Konitow and Bredehoeft, (p. 4) as follows:

1. Darcy's law is valid and hydraulic-head gradients are the only significant driving mechanism for fluid flow.
2. The porosity and hydraulic conductivity of the aquifer are constant with time, and porosity is uniform in space.
3. Gradients of fluid density, viscosity, and temperature do not affect the velocity distribution.
4. No chemical reactions occur that affect the concentration of the solute, the fluid properties, or the aquifer properties.
5. Ionic and molecular diffusion are neglible contributors to the total dispersive flux.
6. Vertical variations in head and concentration are negligible.
7. The aquifer is homogeneous and isotropic with respect to the coefficients of longitudinal and transverse dispersivity.

As was discussed above and is admitted by the model authors, field conditions may violate these assumptions and lead to erroneous model results. Nevertheless, models such as this have been used to give quantitative insights to policy decisions (see Hamilton, 1982; Lewis, 1982). The model users indicate that model results are not necessarily conclusive due to the unknowns associated with model calibration but believe that models are still of use in decision-making.

A number of rather complex partial differential equations represented by finite difference equations are solved in order to calculate the final concentration of pollutants across the grid representing the aquifer. These will not be repeated here. The reader is referred to the USGS's *User's Manual* for a full explanation. We do need to point out that the mathematics involved in this formulation are extremely complex, making it difficult, if not impossible, for the novice user to fully understand all the mathematical and computational underpinnings of this model, and also difficult for some individuals to make full use of the computer model without the expert advice of someone more familiar with this type of work.

A number of different calculations are performed to arrive at an estimate of the of the chemical concentrations:

1. An estimate of the convective transport through each cell based on the movement of a set of tracer particles whose positions change in proportion to the velocity vectors in both the x and y directions.
2. "The changes in concentration caused by hydrodynamic dispersion, fluid sources, divergence of velocity, and changes in saturated thickness are calculated using an explicit finite-difference approximation . . . ." (Konikow and Bredehoft, p. 8)
3. The calculations above must be accompanied by proper specification of boundary and initial conditions. The model can incorporate constant-flux and constant-head conditions to represent well withdrawls or injection, no-flow boundaries, or recharge boundaries. This last condition is represented by a constant-head boundary with a high leakage term. The leakage term is a conductivity constant representing movement of liquid into or out of the aquifer through a confining layer, stream bed, or lake bed. If it is set sufficiently high, the model will calculate a head which is nearly equivalent to a specified constant-head.
4. A mass balance calculation is performed for specified time increments to ensure that there is conservation of mass. In each case, an error is calculated. The error has been found to be less than 8% in several tests of the model versus simple analytical solutions. Lack of data makes more extensive validation infeasible at this point in time.

The computer program is represented by the flow diagram shown as Figure 6-4. Each of the computational steps mentioned above is performed for each time increment. At the end of the period, the

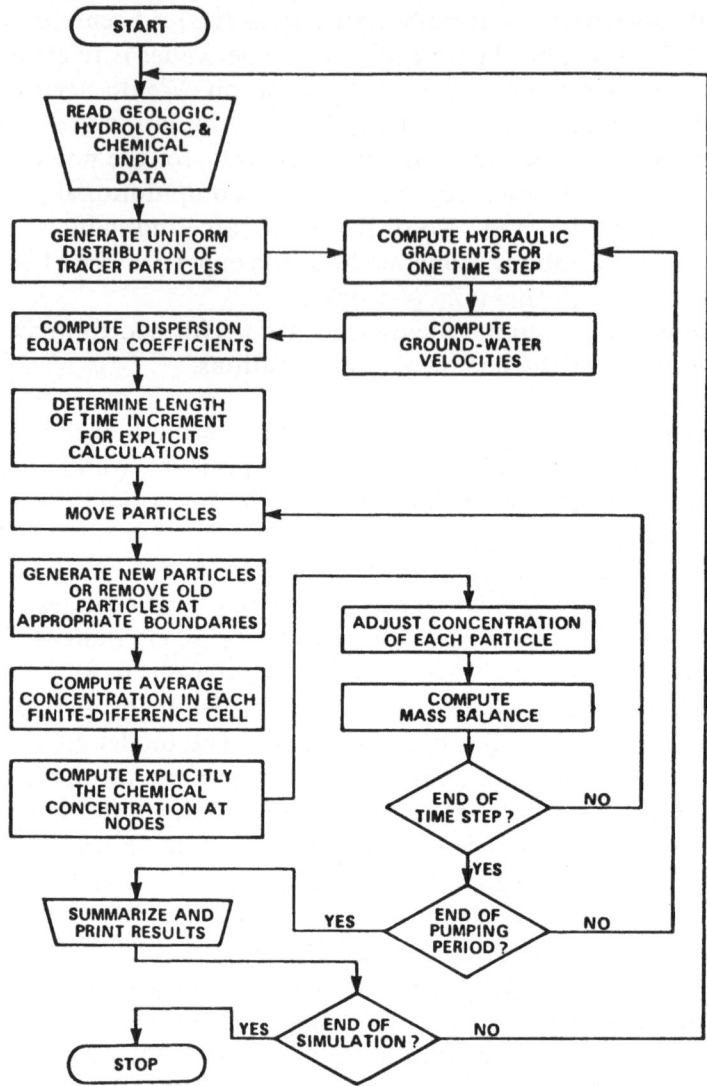

Figure 6-4. Model of solute transport in ground water. (Source: Konikow and Bredehoeft, 1978, p. 21)

results are printed and the run ends. The user may also choose to alter a number of model parameters followed by a new set of simulation runs.

Model data requirements are illustrated by Table 6-7. Here, one can see that there are a large number of parameter specification

Table 6-7. Data Input Formats

Card	Column	Format	Variable	Definition
1	1–80	10A8	TITLE	Description of problem
2	1– 4	I4	NTIM	Maximum number of time steps in a pumping period (limit=100)*.
	5– 8	I4	NPMP	Number of pumping periods. Note that if NPMP>1, then data set 10 must be completed.
	9–12	I4	NX	Number of nodes in x direction (limit=20)*.
	13–16	I4	NY	Number of nodes in y direction (limit=20)*.
	17–20	I4	NPMAX	Maximum number of particles (limit=3200)*. (See eq 71.)
	21–24	I4	NPNT	Time-step interval for printing hydraulic and chemical output data.
	25–28	I4	NITP	Number of iteration parameters (usually 4≤NITP≤7).
	29–32	I4	NUMOBS	Number of observation points to be specified in a following data set (limit=5)*.
	33–36	I4	ITMAX	Maximum allowable number of iterations in ADIP (usually 100 ≤ITMAX≤200).
	37–40	I4	NREC	Number of pumping or injection wells to be specified in a following data set.
	41–44	I4	NPTPND	Initial number of particles per node (options=4, 5, 8, 9).
	45–48	I4	NCODES	Number of node identification codes to be specified in a following data set (limit=10)*.
	49–52	I4	NPNTMV	Particle movement interval (IMOV) for printing chemical output data. (Specify 0 to print only at end of time steps.)
	53–56	I4	NPNTVL	Option for printing computed velocities (0=do not print; 1=print for first time step; 2=print for all time steps).
	57–60	I4	NPNTD	Option for printing computed dispersion equation coefficients (option definition same as for NPNTVL).
	61–64	I4	NPDELC	Option for printing computed changes in concentration (0=do not print; 1=print).
	65–68	I4	NPNCHV	Option to punch velocity data (option definition same as for NPNTVL). When specified, program will punch on unit 7 the velocities at nodes.
3	1– 5	G5.0	PINT	Pumping period in years.
	6–10	G5.0	TOL	Convergence criteria in ADIP (usually TOL≤0.01).
	11–15	G5.0	POROS	Effective porosity.
	16–20	G5.0	BETA	Characteristic length, in feet (=longitudinal dispersivity).
	21–25	G5.0	S	Storage coefficient (set $S=0$ for steady flow problems).

See footnotes at end of table.

Table 6-7. Data Input Formats (cont.)

Card	Column	Format	Variable	Definition
	26–30	G5.0	TIMX	Time increment multiplier for transient flow problems. TIMX is disregarded if $S=0$.
	31–35	G5.0	TINIT	Size of initial time step in seconds. TINIT is disregarded if $S=0$.
	36–40	G5.0	XDEL	Width of finite-difference cell in x direction, in feet.
	41–45	G5.0	YDEL	Width in finite-difference cell in y direction, in feet.
	46–50	G5.0	DLTRAT	Ratio of transverse to longitudinal dispersivity.
	51–55	G5.0	CELDIS	Maximum cell distance per particle move (value between 0 and 1.0).
	56–60	G5.0	ANFCTR	Ratio of T_{yy} to T_{xx}.

Data set	Number of cards	Format	Variable	Definition
1	Value of NUMOBS (limit=5)*	2I2	IXOBS, IYOBS	x and y coordinates of observation points. This data set is eliminated if NUMOBS is specified as =0.
2	Value of NREC	2I2, 2G8.2	IX, IY, REC, CNRECH	x and y coordinates of pumping (+) or injection (−) wells, rate in ft³/s, and if an injection well, the concentration of injected water. This data set is eliminated if NREC=0.
3	a. 1 b. Value of NY (limit=20)*	I1, G10.0 20G4.1	INPUT, FCTR VPRM	Parameter card † for transmissivity. Array for temporary storage of transmissivity data, in ft²/s. For an anisotropic aquifer, read in values of T_{xx} and the program will adjust for anisotropy by multiplying T_{yy} by ANFCTR.
4	a. 1 b. Value of NY (limit=20)*	I1, G10.0 20G3.0	INPUT, FCTR THCK	Parameter card† for THCK. Saturated thickness of aquifer, in feet.
5	a. 1 b. Value of NY (limit=20)*	I1, G10.0 20G4.1	INPUT, FCTR RECH	Parameter card† for RECH. Diffuse recharge (−) or discharge (+), in ft/s.
6	a. 1 b. Value of NY (limit=20)*	I1, G10.0 20I1	INPUT, FCTR NODEID	Parameter card† for NODEID. Node identification matrix (used to define constant-head nodes or other boundary conditions and stresses).
7	Value of NCODES (limit=10)*	I2, 3G10.2, I2	ICODE, FCTR1, FCTR2, FCTR3, OVERRD	Instructions for using NODEID array. When NODEID=ICODE, program sets leakance=FCTR1, CNRECH=FCTR2, and if OVERRD is nonzero, RECH =FCTR3. Set OVERRD=0 to preserve values of RECH specified in data set 5.
8	a. 1 b. Value of NY (limit=20)*	I1, G10.0 20G4.0	INPUT, FCTR WT	Parameter card† for WT. Initial water-table or potentiometric elevation, or constant head in stream or source bed, in feet.
9	a. 1 b. Value of NY (limit=20)*	I1, G10.0 20G4.0	INPUT, FCTR CONC	Parameter card† for CONC. Initial concentration in aquifer.

See footnotes at end of table.

Table 6-7. Data Input Formats (cont.)

Data set	Number of cards	Format	Variable	Definition
10				This data set allows time step parameters, print options, and pumpage data to be revised for each pumping period of the simulation. Data set 10 is only used if NPMP >1. The sequence of cards in data set 10 must be repeated (NPMP −1) times (that is, data set 10 is required for each pumping period after the first).
	a. 1	I1	ICHK	Parameter to check whether any revisions are desired. Set ICHK=1 if data are to be revised, and then complete data set 10b and c. Set ICHK=0 if data are not to be revised for the next pumping period, and skip rest of data set 10.
	b. 1	10I4,3G5.0	NTIM, NPNT, NITP, ITMAX, NREC, NPNTMV, NPNTVL, NPNTD, NPDELC, NPNCHV, PINT, TIMX, TINIT	Thirteen parameters to be revised for next pumping period; the parameters were previously defined in the description of data cards 2 and 3. Only include this card if ICHK=1 in previous part a.
	c. Value of NREC	2I2, 2G8.2	IX, IY, REC, CNRECH	Revision of previously defined data set 2. Include part c only if ICHK=1 in previous part a and if NREC>0 in previous part b.

* These limits can be modified if necessary by changing the corresponding array dimensions in the COMMON statements of the program.

† The parameter card must be the first card of the indicated data sets. It is used to specify whether the parameter is constant and uniform, and can be defined by one value, or whether it varies in space and must be defined at each node. If INPUT=0, the data set has a constant value, which is defined by FCTR. If INPUT=1, the data set is read from cards as described by part b. Then FCTR is a multiplication factor for the values read in the data set.

options, several of which can affect the efficiency and accuracy of the model. For example, the accuracy of representation of the final concentrations changes dramatically with the number of tracer particles per cell specified as NPTPND on the second card, particularly for the earlier periods of calculation. The model authors give some guidance with regard to choosing initial values for these and other parameters but admit that the model has only been tested against rather simplistic analytically solved cases, making its reliability under complex circumstances unknown at best.

Beyond these parametric problems are problems of specification of the aquifer characteristics. These are on card three of the input data. Information on aquifer porosity and dispersivity are most critical and are not widely available in published sources. Similarly, any diversity in values throughout the aquifer, i.e. it is not isotropic, must be specified by the user in following data sets. Given that the data are available, the model is rather straightforward to run and interpret for a simple case and can give some quantitative insights into aquifer behavior with non-reactive chemicals. A number of errors will arise from parameter misspecification, data errors, and assumptions concerning isotropy, chemical concentrations, and chemical reactivity. Until more data are available and there is a better theoretical understanding of the complex processes affecting groundwater movement and chemical transformation, models such as this will have to suffice along with expert judgement and experience.

Chemical Transport and Fate Model (TOXIWASP)

The complexity of the fate of hazardous chemicals also is problematic in surface water environments. Given the number of chemical transformations which might occur, there are both data and analytical problems as is the case with groundwater models. One attempt to account for this complexity has recently been completed by the USEPA (1983). This model is a "dynamic model for simulating the transport and fate of toxic chemicals in water bodies." Both organic chemicals and sediment are simulated by this model. The model attempts to account for the full array of chemical transformations and sediment-chemical exchange processes through a set of transport, chemical transformation, and mass loading submodels.

Figure 6-5 illustrates the nature of the model formulation. The model simulates the concentration of the chemical and suspended sediment and their interaction with each other according to two general equations:

$$\frac{\partial C_1}{\partial t} = u \frac{\partial C_1}{\partial x} + \frac{\partial}{\partial x}\left(E \frac{\partial C_1}{\partial x}\right) + \frac{W_1}{V} - K + \dot{S}_1$$

$$\frac{\partial C_2}{\partial t} = u \frac{\partial C_2}{\partial x} + \frac{\partial}{\partial x}\left(E \frac{\partial C_2}{\partial x}\right) + \frac{W_2}{V} + \dot{S}_2$$

where

C_1 = concentration of chemical, ML^{-3};
C_2 = concentration of suspended sediment, ML;
u = flow velocity of water, LT^{-1};
W_1 = mass loading of chemical, MT^{-1};

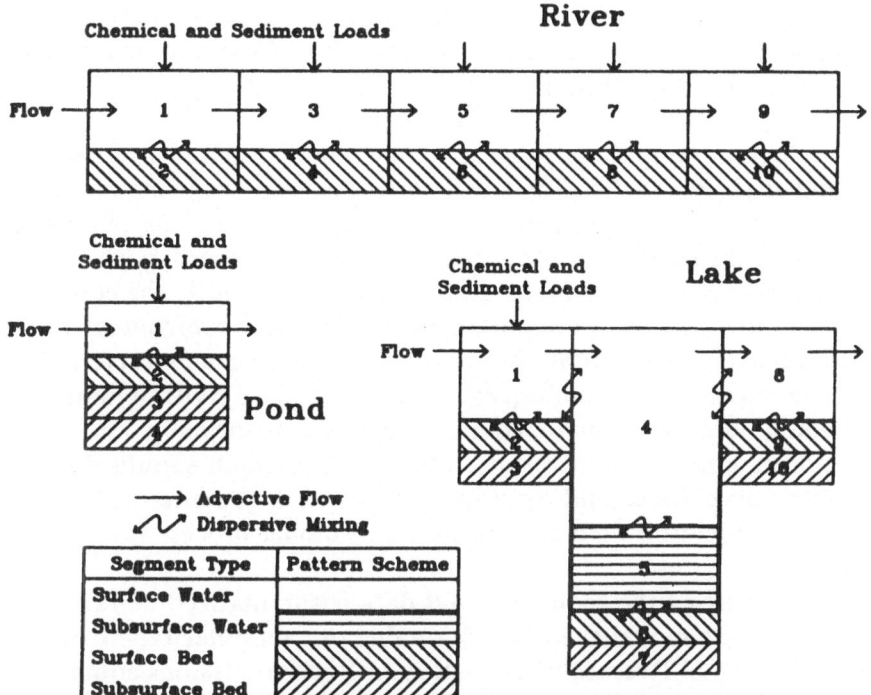

Figure 6-5. Examples of TOXIWASP network configurations. (Source: USEPA, 1983, p. 6)

W_2 = mass loading of sediment, MT^{-1};
$\dot{S}_1$ = net exchange of chemical with bed, $ML^{-3} T^{-1}$;
$\dot{S}_2$ = net exchange of sediment with bed, $ML^{-3} T^{-1}$;
x = longitudinal distance, L;
t = time, T;
M_s = mass of sediment.

TOXIWASP calculates the concentrations for every segment of the modelled water body including surface water, subsurface water, surface bed, and subsurface bed. The model also accounts for advection dispersion, bed sedimentation and erosion, pore water diffusion, hydrolysis, photolysis, oxidation, biodegradation, and volatilization. Although the model provides very complete capabilities, it remains for the user to provide a wide array of coefficients and equations which fully describe each of the processes that are modelled. Some default calculations are built into the model, but all of these equations depend, at a minimum, on rates and other constants supplied by the user. As the forward to the WASP model (which is imbedded in TOXIWASP) says (USEPA, 1983):

> In documenting these programs, our intent is to allow researchers and engineers to develop new theories and apply existing models to their problem. Operating complex, dynamic models should be approached with caution, however, An experienced Fortran programmer should be employed to operate and modify the computer programs to fit the computer and application confronted. Experienced environmental scientists and engineers and preferably a team consisting of biologists, limnologists, and hydrodynamicists would ideally be involved in a program to develop and apply models and interpret results. Modelling research should dovetail with surveillance and experimental research provides calibration and verification data and estimates of model process rates.

The nature of the required input data is summarized in Table 6-8. Although the model is easily obtainable on computer tape from the Environmental Research Laboratory in Fortran versions compatible with both PDP 11/70 and IBM 370 computers, it is not easy for the novice to learn. The documentation of the computer version is extremely terse and does not provide complete information on the options that are available and the impact that option selection will have on computational accuracy. Thus, the user must become familiar with almost all of the potential decisions a priori and then apply the model as a tool. Test run examples are provided on the tape but not in the written documentation. This also makes it difficult for the new user.

The example model run given in the next section further illustrates the potential uses for this model. A complete review of the model equations would duplicate the model documentation and thus will not be made here. What may be of great importance in the future is the growing literature on the nature of the processes which can be modelled using this computer code. As more empirical work is undertaken, it will be possible for users to rely on rates and equations in the literature for the transformation processes (see, for example, USEPA, 1979 and other materials cited in the references at the end of the book). In such circumstances, models such as TOXIWASP can be used with more confidence, than is the case at present.

Table 6-8. Input Data for TOXIWASP

Card
Group

A. Model Identification and System Bypass Options
  1. Model identification numbers
  2. Title card
  3. Simulation option
  4. System bypass option

B. Exchange Coefficients
  1. Number of exchange coefficients, input option number
  2. Scale factor
  3. Exchange coefficients
  4. Exchange bypass options

C. Segment Volumes
  1. Number of volumes, input option number
  2. Scale factor
  3. Volumes

D. Flow
  1. Number of flows, input option number
  2. Scale factor
  3. Flows
  4. Flow bypass option

E. Boundary Conditions
  1. Number of boundary conditions, input option number
  2. Scale factor
  3. Boundary Conditions
  Cards 1-3 are inputted for each system of the model.

F. Forcing Functions
  1. Number of forcing functions, input option number
  2. Scale factor
  3. Forcing functions
  Cards 1-3 are inputted for each system of the model.

G. Parameters (Environmental Characteristics)
  1. Number of parameters
  2. Scale factors
  3. Parameters

H. Constants (Chemical Characteristics and Special Options)
  1. Number of constants
  2. Constants

I. Miscellaneous Time Functions (Environmental Variability)
  1. Number of time functions
  2. Functions name, number of breaks in function
  3. Time function
  Cards 2 and 3 are inputted for each time function required by the model.

J. Initial Conditions
  1. Initial conditions for each system of the model, i.e., chemical and sediment

Table 6-8. Input Data for TOXIWASP (cont.)

Card Group

K. Stability and Accuracy Criteria
   1. Stability criteria
   2. Accuracy criteria

L. Intermediate Print Control
   1. Print interval
   2. Display compartments

M. Integration Control Information
   1. Integration option
   2. Time warp scale factor
   3. Integration interval and total time

N. Display Parameters
   1. Variable names
   2. Dump parameters
   3. Printer plot parameters (time history) cards 1 and 2 are read for each system; etc.
   4. Printer plot parameters (spatial profile)
   5. Pen plot parameters

Source: USEPA, 1983, p. 50–51

USGS Solute Transport Model

Figure 6-6 illustrates a simple example flow problem for the solute transport model. A constant contaminant source is located near the surface of the aquifer, adjacent to a no-flow boundary. The flow field is also surrounded by a no-flow boundary. The flow field consists of two constant-head boundaries with a withdrawl well. The input is represented in numeric form in Table 6-9. When the input data are cross-correlated with the formats shown in Table 6-7, one can see that the problem indicates that a 9 by 10 node system of grids (900 feet square) will be simulated for a period of 2.5 years. Observation wells located at nodes 5,4 and 5,7 are assigned and the model is run using the flow field indicated in Figure 6-6.

The output from the model includes the following items:

1. Recap of the input parameters.
2. A transmissivity map based on the input flow field in the form of a matrix of numbers representing each of the nodes.
3. Diffuse recharge and discharge (in this case set to zero).

Table 6-9. Input Data for Figure 6-6.

```
Card 1   TEST PROBLEM NO. 3 (STEADY FLOW, 1 WELL, CONSTANT-HEAD BOUNDARIES)
Card 2      1   1    9  103200   1   7   2 100   1   9   2  10   1   0    0    0
Card 3      2.5.0001 0.30 100.   0.0   0.0   0.0 900.  900.   0.3 0.50   1.0
```

Data Set 1 {
```
5 4
5 7
```
Data Set 2   `4 7   1.0`

Data Set 3   `0  0.1`

Data Set 4   `0  20.0`

Data Set 5   `0  0.0`

Data Set 6 {
```
1  1.0
000000000
022111220
000000000
000000000
000000000
000000000
000000000
000000000
022222220
000000000
```

Data Set 7 {
```
2   1.0          0.0       0.0     0
1   1.0        100.0       0.0     0
```

Data Set 8 {
```
1  1.0

0.0100.100.100.100.100.100.100. 0.0




0.0 75.  75.  75.  75.  75.  75.  75.  0.0
```

Data Set 9   `0  0.0`

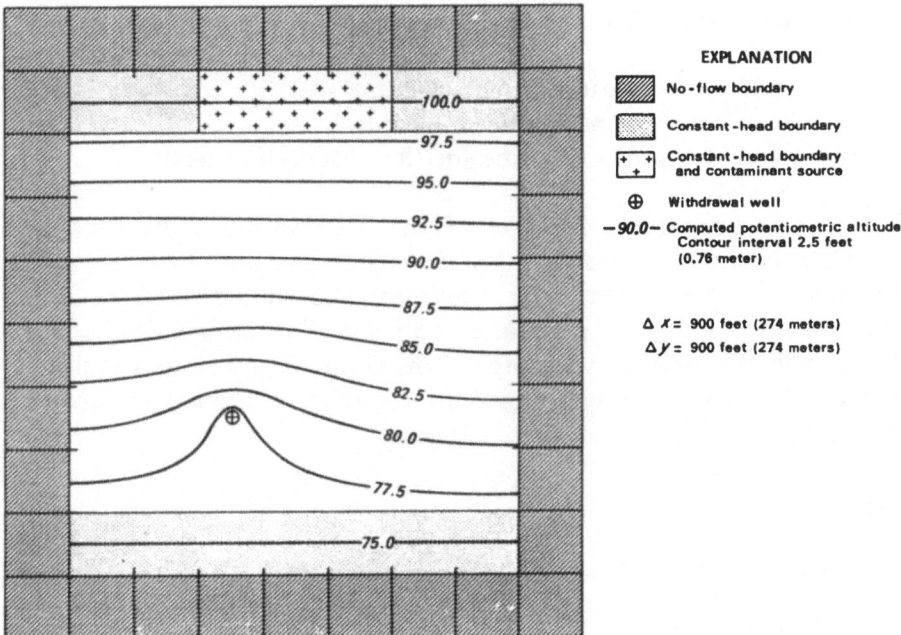

Figure 6-6. Flow problem for solute transport model. (Source: Konitow and Bredehoeft, 1978, p. 32)

183

Table 6-10. Summary Table for Observation Well

OBS.WELL NO.	X	Y	N	HEAD (FT)	CONC.(MG/L)	TIME (YEARS)
1	5	4				
			0	0.0	0.0	0.00
			1	92.0	0.0	0.13
			2	92.0	0.2	0.26
			3	92.0	1.2	0.39
			4	92.0	2.9	0.53
			5	92.0	15.5	0.66
			6	92.0	33.0	0.79
			7	92.0	53.1	0.92
			8	92.0	64.6	1.05
			9	92.0	72.9	1.18
			10	92.0	79.8	1.32
			11	92.0	85.4	1.45
			12	92.0	89.4	1.58
			13	92.0	92.2	1.71
			14	92.0	94.3	1.84
			15	92.0	95.8	1.97
			16	92.0	97.0	2.11
			17	92.0	97.8	2.24
			18	92.0	98.4	2.37
			19	92.0	98.7	2.50

4. A permeability map.
5. For each time iteration:

A map of the transport head;
The drawdown;
Velocities at nodes and on boundaries in both the x and y dimensions;
The final concentrations at the nodes;
The chemical mass balance calculations;
The concentrations over time for the observation wells.

The summary table for one of the observation wells for this problem is shown as Table 6-10. Here, one can see that there is a build-up of chemical over time, beginning at zero and climbing until it reaches a level of 98.7 mg/l after 2.5 years. This information could be used in comparison with other runs simulating different loading rates and flow conditions to assertain the relative impacts of various situations.

REFERENCES

Baski, Henry A. (1979). "Ground-Water Computer Models- Intellectual Toys, *Ground Water,* Vol. 17, No. 2 (March-April), p. 177–179.

Faust, Charles R. and James W. Mercer (1980). "Ground-Water Modeling: Numerical Models," *Ground Water,* Vol. 8., No. 4 (July-August), p. 395–409.

Konikow, L. F. and J. D. Bredehoeft, (1978). "Computer Model of Two Dimensional Solute Transport and Dispersion in Ground Water." In *Techniques of Water-Resources Investigation of the United States Geological Survey*. Chapter C2, Book 7. Washington, DC: U.S. Government. Printing Office.

Hamilton, David A. *Groundwater Management Strategy for Michigan. Groundwater Modeling: Selection, Testing, and Use*. Springfield, VA: NTIS PB83-143206.

Lewis, Barney D. et. al. (1982). *Evaluation of a Predictive Ground-water Solute-Transport Model at the Idaho National Engineering Laboratory*. Springfield, VA: NTIS PB82-204066.

USEPA (1978). *Solid Waste Facts*. Fact Sheet SW-694. Washington, DC Office of Solid Waste, USEPA.

USEPA (1979). "Water-Related Environmental Fate of 129 Priority Pollutants." In Volume I, *Introduction and Technical Background, Metals and Inorganics, Pesticides and PCBs* by Michael A. Callahan et al. EPA-440/4-79-029a. Washington, D.C: U.S. Government Printing Office.

USEPA, (1983). *User's Manual for the Chemical Transport and Fate Model (TOXIWASP)*, Version 1, by Robert B. Ambrose, Jr., Sam I. Hill, Lee Mulkey. EPA-600/3-83-005. Athens, CA: Environmental Research Laboratory, USEPA.

USEPA (1982). *Water Quality Analysis Simulation Program (WASP) and Model Verification Program (MVP)- Documentation* by D. N. DiToro, J. J. Fitzpatrick, and R. V. Thomann. Duluth, MN Office of Research Laboratory, USEPA.

APPENDICES

A. Technical Transfer Problems
B. Classified Bibliography

Appendix A
Technical Transfer Problems

INTRODUCTION

If you are going to use any of the models discussed in this book, the first step will be, in most cases, obtaining a copy of the programs in some readily transferable format. For some of the models this is not possible because the models must be derived empirically by the user (such as the statistical DO model) while others may only be available in the form of a listing that must be tediously reentered into the local computer. For most of the models, it is possible to obtain copies either on computer tape or in the form of a punch card deck. In any event, a number of technical problems may arise that can be very frustrating to the prospective user. The purpose of this Appendix is to introduce the nature of many of these problems and to indicate how to avoid as many of them as possible. Familiarity with these potential bugs may help you avoid a very time-consuming debugging process.

THE TRANSFER OF COMPUTER PROGRAMS

Most of the programs that have been reviewed in this book are available from the governmental agencies which originally sponsored their development. Some of the models can be ordered from the National Technical Information Service in Springfield, Virginia (see Appendix B for ordering instructions) while others must be obtained directly from the sponsoring agency. In either case, the models are most often available on computer tape. You must know several things about the computer on which the tape will be read in order to request the tape in the most easily transferable form. A number of things are important:

1. Make certain that your system has a tape drive and determine the nature of that drive. Most larger systems have tape drives but many smaller ones do not. Obviously, there is no point in ordering a tape if you cannot use it. The nature of the drive also varies. The most flexible units can read several types of tapes. The critical parameters are the following:
    a. *7 or 9 track tapes.* Most newer systems exclusively use 9 track tapes (referring to the number of channels onto which data are recorded)

but some older systems may use 7 track tapes. It is generally possible to specify which type in your order.

b. *Tape labelling.* IBM systems use a standard label which encodes the manner in which the tape is formatted. When the system wants to read the tape, the label is consulted to learn how to do so. Other systems may not be able to read IBM standard labels. IBM systems can read non-labelled tapes but one must have documentation which indicates the tape format information. For the most part, the governmental agencies which reproduce the computer tapes use IBM compatible formats. If you want something else, you must specify it.

c. *Label information.* Each tape that is labelled begins with a volume label. This is the name for the entire tape volume. This label is followed by a file header label containing a description of the data set: record format, blocksize, record size, data set name. Then comes the data set, a tape mark, and a trailer label. Following this comes the header label for the next data set, the data set, and so on, until all the data sets have been passed. The last data set is followed by two tape marks. On a non-labelled tape, none of the label information is recorded. Here, the tape consists of data sets separated by tape marks. On a non-standard label tape, other information, unique to the system, is frequently combined with the standard information. On a non-labelled tape, you must know the order and nature of each file in order to easily read and transfer the information to the host computer.

d. *DCB parameters.* On both labelled and non-labelled tapes, the format of the tape must be known or properly recorded in order to read the tape. In general, tapes with Fortran programs have a card image format; that is, the record size is 80 columns, mimicking an 80 column Hollerith punch card. This information is indicated using the LRECL parameter. Cards are stored on the tape in blocks of data which must be multiples of the record length. The BLKSIZE parameter is used to indicate this value. The type of record format used is related to the nature of the record and block length factors. In general, these types of tapes used fixed or fixed block formats, RECFM=F or FB. The next critical parameter is the density of recording on the tape. This can be 200, 556, 800, 1600, 6250 bytes per inch or DEN=0, 1, 2 3, or 4 respectively. Finally, you must know the method of recording which is indicated by the TRTCH parameter. The default for 9 track tapes is odd parity, translation off, conversion off. For systems with older 7 track drives, these defaults must be changed using various TRTCH options. You must know all of these parameters in order to properly read or copy a tape.

2. After finding out what constraints are imposed by your system, order the

tape with the most compatible format. Consult with the local computer operator if you need help.
3. If the tape cannot be read by your system, commercial translation services are available. This, of course, adds to the cost of the model acquisition.

Even with all these precautions taken, a number of things can go wrong or at least become bothersome problems. Below we list a sequence of possible events and their potential solutions. We begin with the receipt of the tape and end with the making of a copy useable on the local system. Then, we go on to the next set of problems—using the program. The process might go as follows:

1. The tape is received, is not labelled, and does not have complete documentation. Alternatively, there are labels that are not readable by your system or the documentation of the files does not match what is on the tape. In all of these cases, the system will not be able to read the tape. It will try and then give one or more I/O error messages which are completely uninterpretable, even if you know how to look up the message(s) in the local messages and codes documentation. In response, in descending order of ease, you must
   a. *Hire a local expert.* Find someone familiar with tape translation to help out. Give the expert all you know about the tape and what is on it; ask him or her to read the tape and copy it to a new tape easily compatible with the local system.
   b. *Find a local diagnostic program.* Every system has a number of utility programs which attempt to try various ways of reading tapes and comparing what the user expects with what is actually found. For a labelled tape, a tape map indicates the names and DCB parameters for each file on the tape. For non-labelled tapes, the program reads the first several records of each file and describes what it finds. These programs are relatively easy to use but sometimes leave one puzzled by providing output in hexadecimal code. This must be converted to the more familiar alpha-numeric codes before one can see what is going on. For tapes with non-standard labels, this process can be frustrating because the first several records will be labels and not the expected data sets or programs. In such cases, trial and error may be necessary albeit frustrating. Fortunately, most major programs are generally well enough documented to avoid this problem.
   c. *Read a portion of each file and dump it to the printer.* Once you have figured out how the data are formatted, use another utility program to dump a portion of each file to the printer. In this way, you can figure

out whether the file contains Fortran code, example data sets, or other documentation. Once you know all of this, you can go to the next step.

2. You use a local utility program to copy the original tape to a backup, blank tape. The program fails. The potential problems are the following:

   a. *You do not have the proper formats.* Go back to #1.

   b. *The utility program can't handle this format.* Some utility tape-copy programs expect standard label tapes. Make sure the one you use can handle the given situation.

   c. *You did not properly communicate the tape format.* You must use Job Control Language (JCL) to tell the system the format of each tape file. Otherwise, the program assumes some default values which are wrong most of the time. Even if the program is capable of reading and copying this type of tape, you must properly indicate the tape format.

   d. *You hit a system JCL limit.* On tapes with a large number of files (TOXIWASP has 66, for example) the number of JCL instructions may reach a system limit. In this case, multiple jobs must be run to copy the tape, with each job copying a portion of the files.

   e. *You did not properly initialize the blank tape.* The problem may be with the new tape to which you are copying. You must initialize it with a volume name, and indicate to the system you wish to write onto it. Each system has its own conventions for specifying these steps.

3. Once a copy of the tape is made, ensuring that you have a backup if something goes wrong and that you know what you are reading, you are now ready to try to use the model. Hopefully, the user's manual will provide instructions for compiling and running the model. Here, there are two sets of questions which must be answered. First, what special JCL may be required to run the model? Many of the larger models require temporary scratch data sets in order to run properly and adhere to certain conventions with regard to the allocation of input and output data devices. Second, can you asertain how to compile the program and prepare a load module for the local system on which the model is run? Each of these steps has some potential pitfalls:

   a. *The JCL is not well documented.* Each system has its own JCL conventions. Relatively standard for Fortran programs are data set and input/output device allocations. This is done using data set allocations of the form

```
//FT05F001 DD DSN=name, UNIT=disk or tape, VOLUME=location,
          DISP=disposition(new/old, keep/delete)
```

The first part of this statement, FT05, identifies the unit number referred to in the program in a read or write statement. The Fortran statement READ(5,100) means read off unit 5 using format statement 100. In traditional card oriented systems, unit 5 is identified as the card reader but it theoretically can be any disk or tape device, with any volume number and data set name, either new or old, and which can be kept or deleted after the run. By convention, unit 6 is generally the printer (DD SYSOUT=A) and unit 7 is the output punch device (DD SYSOUT=B). Other units up to 42 or more on larger systems, can be allocated to provide for multiple input or output data sets or devices and to provide temporary disk space for intermediate calculation results.

The problem is that for many large programs, such device allocation is the worst part of the documentation. Thus, you must often proceed without all of the data devices being specified in the JCL. When the device number is called in the program, this will lead to a message like "missing FT09F001 dataset," a nasty surprise to the uninitiated. If such occurs, first comb the documentation for a relatively inconspicuous table or one page summary of JCL requirements. If neither exists, there may be an example run, including JCL, on a separate tape file that can be used as a model. If both the foregoing approaches fail, the only possibility is to examine the program and try to figure out the format and space requirements for the missing data set.

b. *The program has a fatal error during compilation.* In order to compile a Fortran program, you must send the fortran code to the compiler on the local computer. Unfortunately, there are many versions of Fortran compilers for different systems or on the same type of systems of different ages. There are at least six major, different Fortran compilers for IBM equipment alone. Depending on which compiler was used to develop the program you are attempting to use and how different the syntax is from your version, a number of compilation errors may occur. If these are severe or numerous enough the program terminates before running. The only solutions are to try a different compiler if available, or to edit the syntax to conform to the local conventions, or both. Here, systems with on-line editor programs are very advantageous, allowing the knowledgable user to transfer the code to disk and edit it using the computer. However, this task is time consuming if you do not already know how to use the editor. A less sophisticated alternative is to dump the program to cards, replace the erroneous cards, and read the revised program back into the system.

4. Now that we have gotten through compilation, the next step is to create a load module of the program. This is a version of the program in machine language or directly executable form. This step is not required but is desirable since a load module costs much less to run when it avoids a

compilation step and takes less time. Each compiler should have a built-in link edit step. By consulting with the local computer operator, you should be able to get the necessary JCL to save the load module on disk for further use.

5. Now you are ready to go. In order to help in model transfer, almost all program tapes include an example input data set. Once you have a program that will compile or a load module, you need only submit the example data set in the proper place in the input sequence and (finally) you should get some output. Still, something could go wrong:

   a. *Input data set errors.* Most input example data sets are correct but sometimes have quirks which seem unimportant but are critical to the model running correctly. For example, CDMQC requires strategically located blank cards representing the end of each major data set. The uninitiated user may think these extraneous and remove them from the deck. You must study the required input card sequence carefully and look for such potential pitfalls.

It is obvious that a large number of technical problems can occur during model transfer. For this reason, you must realistically plan to spend a significant amount of time to debug and make the initial model runs. Additional time will be required to become familiar and comfortable with input requirements, the interpretation of results, and so forth. Whenever possible, personnel with both computer and modelling expertise should be available to help with the model tranfer. It is still possible for the relatively uninitiated to use such programs, but a longer and perhaps more difficult debugging process is likely to be required.

Appendix B
Classified Bibliography

INTRODUCTION

Aside from the models that are reviewed in this book, a large number of models are available at nominal cost. These are supported, in turn, by a rather extensive literature about modelling, various methodological approaches to modelling, and applications concerning models. This bibliography is an attempt to reference a good portion of this broad literature. In order to make them more useful, the references are classified by subject area. In organization, the first level of classification follows the chapters of this book. Thus, we first cite introductory materials then studies of water quality, stormwater, air quality, and so forth. Under each first level heading, the second level of classification involves the nature of the material. Here we have created four major subject categories:

1. Overview material;
2. Reviews of models and methods;
3. User's guides and model documentation;
4. Instructions to obtain computer models on tape or cards.

Although the categories overlap somewhat forcing an arbitrary classification at times, we believe that the bibliography will be a useful resource to the professional attempting to become familiar with a particular area.

One other problem should be mentioned. Many of the technical governmental documents are generally available free of charge from the originating agency while their limited supply lasts. After that time, the documents must be purchased through the National Technical Information Service, U.S. Department of Commerce, 5285 Port Royal Road, Springfield, VA 22161. (telephone (703) 487-4660; Orders must be prepaid.) Documents are available in microfiche for $4.50 or in paper, where the cost is proportional to the document length. In the Bibliography, these technical documents are sometimes referred to by the original agency number (EPA for example) and sometimes by an NTIS number (these start with PB, ANL, CONF, and so on). One can order the documents from NTIS using either number; and

, the relevant agency can be contacted to determine if any free copies are still available. Agency addresses should be available in major libraries.

MODELS IN ENVIRONMENTAL PLANNING

General Background Materials

Bertalanffy, L. V. (1968). *General Systems Theory.* New York: George Brazilier.

Canter, Larry W. (1977). *Environmental Impact Assessment.* New York: McGraw-Hill Book Co.

Fisch, Oscar (1980). "Spatial Equilibrium with Locational Interdependencies—The Case of Environmental Spillovers," *Regional Science and Urban Economics,* Vol. 10, p. 201–209.

Haimes, Yacov Y. (1982). "Modeling of Large Scale Systems in a Hierarchical Multiobjective Framework" in *Large Scale Systems,* ed. by Yacov Y. Haimes. New York: North Holland Publishing Co. p. 1–17.

Jain, Ravinder K. and Bruce L. Hutchings, eds. (1978). *Environmental Impact Analysis: Emerging Issues In Planning.* Chicago: University of Illinois Press.

Punnett, Richard E. (1982). *District Considerations in Mathematical Modelling.* Springfield, VA: NTIS AD-P000413/5.

Sewell. Granville H. (1975). *Environmental Quality Management.* Englewood Cliffs: Prentice-Hall.

U.S. Environmental Protection Agency (1983). *A Fortran Program for Computing the Pollutant Standards Index.* Springfield, VA: NTIS PB83-154757.

Walker, Warren E. (1982). *Models in the Policy Process: Past, Present, and Future.* Springfield, VA: NTIS AD-A 113334/7.

Models and Methods

Andrews, John F. (1978). "Applications of Some Systems Engineering Concepts and Tools to Water Pollution Control Systems," in *Mathematical Models in Water Pollution Control,* ed. by A. James. New York: John Wiley & Sons.

Chankong, V. and Y. Y. Haimes (1983). *Multiobjective Decision Making: Theory and Methodology.* New York: Elsevier-North Holland.

Fisch, Oscar and Steven I. Gordon (1980). *Evaluation and Testing of NSF-RANN Sponsored Land-Use Modeling Projects with Ohio as a Test Case.* 7 vols. Columbus, OH: Ohio State University Research Foundation.

Frendewey, J. O., et al. (1977). *General Guidelines Regarding the Transferability of Computer-Based Socio-Economic, Land Use, and Environmental Models.* Springfield, VA: NTIS PB-275254/IGA.

Jameson, David L. (1976). *Ecosystem Impacts of Urbanization Assessment Methodology.* Springfield, VA: PB-259931.

Thomann, Robert V. (1982). "Verification of Water Quality Models," *Journal of the Environmental Engineering Division, Proceedings of the American Society of Civil Engineers,* Vol. 108, No. EE5, p. 923–940.

Timmons, John F. (1982). *Development and Application of Models for Analyzing Water Resource Use and Development Within a Regional Framework of Economic Growth and Environmental Quality.* Springfield, VA: NTIS PB83-167932.

U.S. Environmental Protection Agency (1983). *Use of Models for Water Resources Management, Planning, and Policy.* Springfield, VA: NTIS PB83-103655.

Werner, Kirk G., et al. (1976). *Asbestos and Silicate Pollution (citations from the NTIS Data Base).* Springfield, VA: NTIS PS-76/0981.

Willey, R. (1982). *River and Reservoir Systems Water Quality Modeling Capability.* Springfield, VA: NTIS AD-A114192/8.

WATER QUALITY MODELS

General Background

Bartsch, A. F. (1972). *Role of Phosphorous in Eutrophication.* Springfield, VA: NTIS PB-228292.

Butz, B. P., B. A. G. Lewis, H. P. Friesma, and A. N. Holter (1975). *Proceedings of Workshop on the Social Consequences of Waste Heat Discharge Alternatives.* Springfield, VA: NTIS CONF-750692.

Chen, Charles, W., et al. (1972). *Ecologic Simulation for Aquatic Environments.* Springfield, VA: NTIS PB-218828.

Davis, Michael J. (1981). *Conceptual Design of a Regional Water Quality Screening Model.* Springfield, VA: NTIS DOE/EV 10154/T1.

Davis, Michael J. et al. (1982). *River Basin Validation of the Water Quality Assessment Methodology for Screening Nondesignated 208 Areas.* Vol. I: *Nonpoint Source Load Estimation.* Springfield, VA: NTIS PB82-260837.

Dean, J. David et al. (1982). *River Basin Validation of the Water Quality Assessment Methodology for Screening Nondesignated 208 Areas.* Vol. II: *Chesapeake-Sandusky Nondesignated 208 Screening Methodology Demonstration.* Springfield, VA: NTIS PB82-260845.

Fair, Gordon Maskew, John Carles Geyer, and Daniel Alexander Okun (1971). *Elements of Water Supply and Wastewater Disposal.* New York: John Wiley & Sons.

Field, Richard et al. (1977). *Urban Runoff Pollution Control Technology Overview.* Springfield, VA: NTIS PB-264452.

Formica, Peter N. (1976). *Controlled and Uncontrolled Emission Rates and Applicable Limitations for Eighty Processes.* Springfield, VA: NTIS PB-266978.

Frea, James I. et al. (1972). *Washout Processes in Lake Systems.* Springfield, VA: NTIS PB-217884.

Hammer, Mark J. and Kenneth A. Mackichan (1981). *Hydrology And Quality of Water Resources.* New York: John Wiley & Sons.

Kemp, Homer T., Robert L. Little, Verno L. Holoman, and Ralph L. Darby (1973). *Water Quality Data Book.* Vol. 5: *Effects of Chemicals on Aquatic Life.* Washington, DC: USEPA 18050HLA.

Martin, Donald M. et al. (1972). *The Role of Nitrogen in the Aquatic Environment.* Springfield, VA: NTIS PB-213496.

Murphy, Jeanne S. et al. (1977). *Municipal Water Pollution Control Abstracts.* Springfield, VA: NTIS PB-264663.

Oberts, Gary L. (1982). *Water Resources Management: Non-Point Source Pollution Technical Report, Final Report, 1980.* Springfield, VA: NTIS PB82-225087.

Office of Water Research and Technology (1976). *Agricultural Runoff, A Bibliography,* Volume 2. Springfield, VA: NTIS PB-255849.

Rumker, R. V. et al. (1972). *The Use of Pesticides in Suburban Homes and Gardens and Their Impact on the Aquatic Environment.* Springfield, VA: NTIS PB-213960.

Sargent, Frederick O. (1976). *Land Use Patterns, Eutrophication, and Pollution in Selected Lakes.* Springfield, VA: NTIS PB-263501.

Smith, D. E. (1973). *Relative Leaching Rates of Common Nitrogen Carriers from Sandy Soils in Relation to Lake Eutrophication.* Springfield, VA: NTIS PB-225474/6.

Streeter, H. W. and E. B. Phelps (1925). "A Study of the Pollution and Natural Purification of the Ohio River," Bulletin No. 146, U.S. Public Health Service.

U.S. Environmental Protection Agency (1983). *Approach to Solving a Basin Wide Water Resources Management Planning Problem with Multiple Objectives.* Springfield, VA: NTIS PB83-150052.

U.S. Environmental Protection Agency (1976). *Areawide Assessment Procedures Manual.* 3 vols. Cincinnati, OH: Municipal Environmental Research Laboratory, U.S. Environmental Protection Agency EPA-600/9-76-014.

U.S. Environmental Protection Agency (1983). *Planning Guide for Evaluating Agricultural Non-Point Source Water Quality Controls.* Springfield, VA: NTIS PB83-119453.

U.S. Environmental Protection Agency (1983). *Urban Runoff Quality: Information Needs.* Springfield, VA: NTIS PB83-117002.

URS Research Co. (1974). *Water Quality Management Planning for Urban Runoff.* Springfield, VA: NTIS PB-241689.

Zison, Stanley W. et al. (1977). *Water Quality Assessment: A Screening Method for Nondesignated 208 Areas.* Springfield, VA: NTIS PB-277161.

Models and Methods

American Public Works Association (1969). *Water Pollution Aspects of Urban Runoff.* Springfield, VA: NTIS PB-215532.

Amy, Gary, Robert Pitt, Rameshawar Singh, Westly Bradford, and Michael LaGraff (1974). *Water Quality Management Planning for Urban Runoff.* Springfield, VA: NTIS EPA-440/9-75-004.

Beck, M. B. (1978). "Modelling of Dissolved Oxygen in a Non-Tidal Stream" in *Mathematical Models in Water Pollution Control,* ed. by A. James. New York: John Wiley & Sons.

Carey, G. W., L. Zobler, M. R. Greenberg and R. M. Hordon (1972). *Urbanization, Water Pollution and Public Policy.* New Brunswick, NJ: Center for Urban Policy Research, Rutgers University.

Characklis, William G., Frank J. Gaudet, Frank L. Roe, and Philip B. Bedient (1979). *Maximum Utilization of Water Resources in a Planned Community— Stormwater Runoff Quality Data Collection Reduction and Analysis.* Springfield, VA: NTIS EPA-600/2-79-050b.

Chui, Tai Wik David et al. (1981). *Highway Runoff in Washington State: Model Validation and Statistical Analysis.* Springfield, VA: NTIS PB83-171520.

Clark, Leo J. et al. (1978). *A Water Quality Modelling Study of the Delaware Estuary.* Springfield, VA: NTIS PB-282984.

Colston, Newton V. Jr. (1974). *Characterization and Treatment of Urban Land Runoff.* Springfield, VA: NTIS EPA-670/2-74-096.

Crosswhite, William M. et al. (1982). *Water Quality Monitoring and Modeling Workshop: Proceedings.* Springfield, VA: NTIS PB82-168394.

Dean, David J. et al. (1982). *River Basin Validation of the Water Quality Assessment Methodology for Screening Nondesignated 208 Areas.* Springfield, VA: NTIS PB82-260845.

Digiano, F. A. (1974). *Definition of Procedures for Study of River Pollution by Non-Point Urban Sources.* Springfield, VA: NTIS PB-237972.

Digiano, Francis A. (1971). *Mathematical Modeling of Nutrient-Transport.* Springfield, VA: NTIS PB-225127/0.

Donigian, Anthony S. Jr. (1976). *Modeling Nonpoint Pollution from the Land Surface.* Springfield, VA: NTIS PB-257089.

Falco, J. W. et al. (1982). *Screening Procedure for Assessing the Transport and Degradation of Solid Constituents in Subsurface and Surface Waters.* Springfield, VA: NTIS PB83-117036.

Gordon, Steven I. and Richard K. Fromuth (1980–81). "A Point, Non-point Source Model of Disssolved Oxygen for the Great Miami River". *Journal of Environmental Systems,* Vol. 10, No. 3, p. 185–200.

Haith, D. A. (1976). "Land Use and Water Quality in New York, River," *Journal of the Environmental Engineering Division, Proceedings of the American Society of Civil Engineers,* Vol. 102, No. EE1 p. 1–28.

Hughto, R. J. and D. P. Loucks (1977). *A Multiparameter Water Quality Management Model for River Basins.* Springfield, VA: NTIS PB-275588/2GA.

Jennings, Marshal E. et al. (1976). *Determination of Biochemical Oxygen Demand Parameters.* Springfield, VA: NTIS PB-253739.

Keefer, Thomas N. et al. (1979). *Dissolved Oxygen Impact from Urban Storm Runoff.* Cincinnati, OH: U.S. Environmental Protection Agency EPA-600/2-79-156.

Law, James A. Jr. et al. (1969). *The Impact of Agricultural Pollutants on Subsequent Users.* Springfield, VA: NTIS PB-215117.

Little, Arthur D. Inc. (1977). *Development of Additional Hazard Assessment Models.* Springfield, VA: NTIS AD-A042365.

Meinholz, Thomas L., William A. Kreutzberger, Martin E. Harper, and Kevin J. Fay (1979). *Verification of the Water Quality Impacts of Combined Sewer Overflow.* Springfield, VA: NTIS EPA-600/2-79-155.

Mills, W. B. et al. (1982). *Water Quality Assessment: A Screening Procedure for Toxic and Conventional Pollutants.* Athens, GA: U.S. Environmental Protection Agency EPA-600/6-82-004a.

Newlin, Joseph T. et al. (1974). *An Analysis of Non-Point Source Pollution in the Rocky Mountain Prairie Region.* Springfield, VA: NTIS PB-255273.

Novotny, Vladimir and Gordon Chesters (1981). *Handbook of Non-Point Pollution, Sources and Management.* New York: Van Nostrand Reinhold.

Pisano, William C., Gerald L. Aronson, Celso S. Queiroz (1979). *Dry-Weather Deposition and Flushing for Combined Sewer Overflows Pollution Control.* Springfield, VA: NTIS EPA-600/2-79-133.

Pitt, Robert (1979). *Demonstration of Nonpoint Pollution Abatement through Improved Street Cleaning Practices.* Springfield, VA: NTIS EPA-600/2-79-161.

Porcella, D. B. et al. (1974). *Comprehensive Management of Phosphorous Water Pollution.* Springfield, VA: NTIS PB-232958.

Rossman, Lewis A. (1979). *Computer Aided Synthesis of Wastewater Treatment and Sludge Disposal Systems.* Cincinnati, OH: U.S. Environmental Protection Agency EPA-600/2-79-158.

Scott, R. Lennie et al. (1975). *Inactive and Abandoned Underground Mines: Water Pollution Prevention and Control.* Springfield, VA: NTIS PB-258263.

Shaheen, Donald G. (1975). *Contributions of Urban Roadway Usage to Water Pollution.* Springfield, VA: NTIS EPA-600/2-75-004.

Shirazi, Mustafa A. and Lorin R. Davis (1974). *Workbook of Thermal Plume Prediction.* Vol. II: *Surface Discharge.* Springfield, VA: NTIS EPA R2-72-005b.

Sylvester, Robert O. and Foppe B. Dewalle (1972). *Character and Significance of Highway Runoff Waters.* Seattle: Department of Civil Engineering, University of Washington. Report for Washington State Highway Commission.

Thomann, R. V. (1967). "Time Series Analysis of Water Quality Data," *Journal of the Sanitary Engineering Division, Proceedings of the American Society of Civil Engineers,* Vol. 93 (February), p. 1–23.

Thomann, Robert V. et al. (1975). *Mathematical Modeling of Phytoplankton in Lake Ontario: Model Development and Verification.* Washington, DC: U.S. Environmental Protection Agency EPA-600/3-75-005.

U.S. Environmental Protection Agency (1983). *An Evaluation of Three Pesticide Runoff Loading Models.* Springfield, VA: NTIS PB83-130047.

Uttormark, Paul D. et al. (1974). *Estimating Nutrient Loadings of Lakes from Non-Point Sources.* Springfield, VA: NTIS EPA-660/3-74-020.

Wang, Multao, Lawrence K. Wang, Jao-Fuan Kao, Ching-Gung Wen, and David Vielkind (1974). "Computer-Aided Stream Pollution Control and Management, Part I," *Journal of Environmental Management,* Vol. 9, p. 165–183.

Wun-Chen Wsng et al. (1973). *A Technique for Evaluating Algal Growth Potential in Illinois Surface Waters.* Springfield, VA: NTIS PB-218955.

Zisson, Stanley W. et al. (1978). *Rates, Constants, and Kinetic Formulations in Surface Water Quality Modeling.* Athens, GA: U.S. Environmental Protection Agency EPA-600/3-78-105.

User's Manuals for Computer Models

Ambrose, Robert B. Jr. et al. (1980). *User's Manual for the Dynamic (Delaware) Estuary Model.* Springfield, VA: NTIS PB81-106742.

Keller, E. C. Jr. et al. (1974). *A Diversity Indices Computer Program for Use in Aquatic Systems Evaluation.* Springfield, VA: NTIS PB-235259.

Litwin, Y. J. et al. (1981). *Areawide Stormwater Pollution Analysis with the Acreoscopic Planning (ABMAC) Model.* Part 1: *Documentation and Application.* Part II: *User's Manual.* Springfield, VA: NTIS PB82-107947.

Medina, Miguel A. Jr. (1979). *LEVEL III: Receiving Water Quality Modeling for Urban Stormwater Management.* Springfield, VA: NTIS EPA-600/2-79-100.

U.S. Environmental Protection Agency (1983). *ANSWERS (Aereal Non-Point Source Watershed Environmental Response Simulation) User's Manual.* Springfield, VA: NTIS PB83-115436.

Computer Program Acquisition

The Delaware Estuary model cited in the previous section is available from the Delaware River Basin Commission, 25 State Police Rd. W., Trenton, NJ 08628. A nominal fee will be charged for copying the tape.

Hughto, Richard J. and Robert P. Schrieber (1982). "Microcomputer water quality simulation model," *Civil Engineering Magazine* (March), p. 58–59. Also see corrections in letters section of April issue.

STORMWATER RUNOFF MODELS

General Background

Barnwell, Thomas O. (1983). *Proceedings of Stormwater and Water Quality Management Modeling Users Group Meeting 25–26, March, 1982.* Springfield, VA: NTIS PB83-145540.

Greenberg, Michael R. and Robert M. Hordon (1976). *Water Supply Planning: A Case Study and Systems Analysis.* New Brunswick, NJ: Center for Urban Policy Research, Rutgers University.

Heaney, J. P. (1977). *Nationwide Evaluation of Combined Sewer Overflows and Urban Storm Water Discharges.* Springfield, VA: NTIS PB-266005/8GA.

Hjelmfelt, A. T. Jr. and J. J. Cassidy (1975). *Hydrology for Engineers And Planners,* Ames, IA: Iowa State University Press.

Horton, R. E. (1945). "Erosional Development of Streams and Their Drainage Basins: Hydrophysical Approach to Quantitative Morphology," *Bulletin of the Geological Society of America,* Vol. 56, p. 275–370.

Knapp, G. L. et al. (1972). *Urban Hydrology—A Selected Bibliography with Abstracts.* Springfield, VA: NTIS PB-219105.

Lazaro, Timothy R. (1979). *Urban Hydrology.* Ann Arbor, MI: Ann Arbor Science Publishers.

Linsley, Ray K. Jr., Max A. Kohler, and Joseph L. H. Paulhus (1975). *Hydrology for Engineers.* New York: McGraw Hill Book Co.

Maloney, Frank E., Richard G. Harmann, and Brian D. E. Carter (1980). "Stormwater Runoff Control: A Model Ordinance for Meeting Local Water

Quality Management Needs," *Natural Resources Journal,* Vol. 20, No. 4, p. 713-764.

McPherson, M. B. (1972). *Hydrologic Effects of Urbanization in the United States.* Springfield, VA: NTIS PB-212579.

McPherson, M. B. et al. (1974). *Management of Urban Storm Runoff.* Springfield, VA: NTIS PB-234316.

Meta Systems, Inc. (1975). *Systems Analysis In Water Resources Planning.* New York: Water Information Center, Inc.

Urban Land Institute (1975). *Residential Stormwater Management.* Washington, DC: Urban Land Institute.

Wanielista, Martin P. (1978). *Stormwater Management (Quantity and Quality).* Ann Arbor, MI: Ann Arbor Science Publishers.

Ward, R. C. (1967). *Principles of Hydrology.* New York: McGraw Hill Book Co.

Yevjevich, Vujica (1975). Introduction to *Unsteady Flow in Open Channels,* Vol. I, ed. by K. Mahmood and V. Yevjevich. Fort Collins, CO: Water Resources Publications, p. 1-28.

Models and Methods

Brown, J. W. et al. (1974). *Models and Methods Applicable to Corps of Engineers Urban Studies.* Springfield VA: NTIS AD-786516.

Driscoll, E. D., et al. (1979). *A Statistical Method for Assessment of Urban Storm Water.* Springfield, VA: NTIS PB-299185/9GA.

Espey, W. H. and D. E. Winslow (1975). *Quantity Aspects of Urban Stormwater Runoff,* Short Course Proceedings, Applications of Stormwater Management Models, National Environmental Research Center, Office of Research and Development. Cincinnati, OH: U.S. Environmental Protection Agency, 1975.

Henderson, F. M. (1963). "Some Properties of the Unit Hydrograph," *Journal of Geophysical Research,* Vol. 68, p. 4785-4793.

Lunard, William G., John Finnemore, Joseph H. Loop, and Robert M. Finn (1980). *Urban Stormwater Management and Technology: Case History.* Springfield, VA: NTIS EPA-600/8-80-035.

Miller, William A. and Jean A. Cunge (1975). "Simplified Equations for Unsteady Flow," in *Unsteady Flow in Open Channels,* Vol. I, ed. by K. Mahmood and V. Yevjevich. Fort Collins, CO: Water Resources Publications, 1975. P. 183-258.

Ohio Environmental Protection Agency, Office of the Planning Coordinator (1982). *Planning and Engineering Data Management System for Ohio: Urban Stormwater Analysis—A Computer Based Methodology.* Columbus. OH.

Overton, Donald E. and Michael E. Meadows (1976). *Stormwater Modeling.* New York: Academic Press.

Sandoski, Dorothy A. et al. (1972). *Selected Urban Storm Water Runoff Abstracts, July 1971-June 1972.* Springfield, VA: NTIS PB-2144122.

Sherman, L. K. (1932). "Streamflow from Rainfall by the Unit-Graph Method," *Engineering News Record,* Vol. 108 (April 7).

U.S. Department of Agriculture (1965). *Rainfall-Erosion Losses from Cropland East of the Rocky Mountains,* Agricultural Handbook 202. Washington, DC: U.S. Government Printing Office.

User's Manuals

Huber, Wayne et al. (1975). *Storm Water Management Model User's Manual,* Volume II. Washington, DC: U.S. Environmental Protection Agency EPA-670/2-75-017.

Lager, John A. et al. (1971). *Storm Water Management Model. Vol. I: Final Report.* Springfield, VA: NTIS PB-203289.

Lager, John A. et al. (1971) *Storm Water Management Model,* Vol. III: *User's Manual.* Springfield, VA: NTIS PB-203291.

Lager, John A. et al. (1971). *Storm Water Management Model* Vol. IV: *Program Listing.* Springfield, VA: NTIS PB-203292.

Lager, John A., Theodor Didriksson, and George B. Otte, (1976). *Development and Application of a Simplified Stormwater Management Model.* Springfield, VA: NTIS EPA-600/2-76-218.

Litwin, Yorum J. et al. (1982). *Areawide Stormwater Pollution Analysis with the Macroscopic Planning (ABMAC) Model.* Part I: *Documentation and Application.* Part II: *User's Manual.* Springfield, VA: NTIS PB82-107947.

U.S. Army Corps of Engineers, Hydrologic Engineering Center (1975). *Storage, Treatment, Overflow, Runoff Model "STORM" User's Manual.* 723-58-17520. Davis, CA: HEC, U.S. Army Corps of Engineers.

U.S. Department of Agriculture, Soil Conservation Service (1965). *Computer Program for Project Formulation Hydrology.* Technical Release 20. Washington, D.C.: USDA.

U.S. Department of Agriculture, Soil Conservation Service (1972). *SCS National Engineering Handbook,* Section 4: *Hydrology."* Washington, DC: U.S. Government Printing Office.

U.S. Department of Agriculture, Soil Conservation Service (1975). *Urban Hydrology for Small Watersheds.* Technical Release 55. Washington, DC.: USDA, 1975.

Computer Program Acquisition

The stormwater models discussed in Chapter 3 are available from the agencies which developed them: TR-20, SCS at nominal charge; STORM from U.S. Army Corps of Engineers at nominal charge; SWMM through EPA, $150.

AIR POLLUTION MODELS

General Background

Battelle Memorial Institute (1974). *The Cost of Clean Air.* Springfield, VA: NTIS PB-238762.

Brewer, A., E. R. Rensberg, and G. E. Woodbury. *A Diagnostic Model for Studying Daytime Urban Air Quality Trends.* Springfield, VA: NTIS N81-22587/2.

Colucci, A. V. (1976). *Sulfur Oxides; Current Status of Knowledge.* Springfield, VA: NTIS EPRI-EA-316.

Datronic Systems Corporation (1973). *Bibliography on Air Pollution Forecasting by Computer and Diffusion Models.* Springfield, VA: NTIS PB-233184.

Guldmann, J. M., and D. Shefer (1980). *Industrial Location and Air Quality Control—A Planning Approach.* New York: John Wiley & Sons.

Guldmann, Jean-Michel (1983). "A Structural Framework for the Design of Integrated Environmental and Land-Use Planning Optimization Models." Paper presented at the 23rd European Congress of the Regional Science Association, Poitiers, France, August 30–September 2.

Lave, Lester B. and Gilbert S. Omenn (1981). *Clearing the Air: Reforming the Clean Air Act.* Washington, DC: The Brookings Institution.

Miller, C. W. and D. E. Fields (1982). *Applicability of Gaussian Plume Dispersion Parameters to Acute Radionuclide Releases.* Springfield, VA: NTIS CONF-801107-44.

Miller, Catherine G. (1981). *Case Studies in the Application of Air Quality Modeling in Environmental Decision Making.* Springfield, VA: NTIS PB81-213233.

Stern, Arthur C., ed. (1976). *Air Pollution,* 3rd Ed. 7 vols. New York: Academic Press.

U.S. Environmental Protection Agency (1977). *Coal Cleaning with Scrubbing for Sulfur Control: An Engineering Economic Summary.* Washington, DC: Decision Series, USEPA, EPA-600/9-77-107.

U.S. Department of Health, Education, and Welfare (1969). *The Cost of Clean Air.* Springfield, VA: NTIS PB-257587.

U.S. Environmental Protection Agency (1973). *The Cost of Clean Air.* Springfield, VA: NTIS PB-257949.

U.S. Environmental Protection Agency (1983). *Cost Analysis of Proposed Changes to the Air Quality Modeling Guidelines.* Springfield, VA: NTIS PB83-112177.

U.S. Environmental Protection Agency (1974), *A Guide for Considering Air Quality in Urban Planning.* Research Triangle Park, NC: EPA-450/3-74-020.

U.S. Environmental Protection Agency (1983). *Regional Workshops on Air Quality Modeling: A Summary Report.* Springfield, VA: NTIS PB83-150573.

Models and Methods

Beaton, John L. et al. (1972). *Air Quality Manual.* Vol. 1: *Meteorology and Its Influence on the Dispersion of Pollutants from Highway Line Sources.* Springfield, VA: NTIS PB-219811.

Cirillo, Richard and George Concaildi (1982). *Development of Computerized Emission Projection and Allocation System Phase II: Comparison of Existing Systems, Final Report.* Springfield, VA: NTIS PB82-240276.

Cleveland, Jerry G. et al. (1970). *Evaluation of Dispersed Pollutional Loads from Urban Areas.* Springfield, VA: NTIS PB-203746.

Darling, E. M. Jr. et al. (1974). *Computer Analysis of Air Pollution from Highways, Streets, and Complex Interchanges: A Case Study.* Springfield, VA: NTIS PB-231334.

Emmanuel, W. R. et al. (1977). *Optimization Model for Air Quality/Analysis in Energy Facility Siting.* Springfield, VA: NTIS ORNL/TM-6007.

Fishelson, G., G. Rausser, and A. Cohen (1976). *Air Pollution and the Siting of Fossil Fuel Power Plants.* Springfield, VA: NTIS ANL-76-xx-14.

Gipson, Gerard L. (1982). *Comparison of Three Ozone Models: Urban Airshed, City-Specific EKMA, and Proportional Rollback, Final Report.* Springfield, VA: NTIS PB82-234089.

Grumman Aerospace Corporation (1972). *A Survey of Dispersion Coefficients for Estimating Pollution Transport.* Springfield, VA: NTIS AD-754010.

Gustafson, S. A. et al. (1976). *Numerical Optimization Techniques in Air Quality Modeling. Objective Interpolation Formulae.* Springfield, VA: NTIS PB-262200.

Hanna, S. R. (1981). *Handbook on Atmospheric Diffusion Models.* Springfield, VA: NTIS ATDL-81/5.

Hillyer, Martin S. (1982). *Evaluating Simple Oxidant Predictions Methods Using Complex Photochemical Models: Cluster Analysis Applied to Urban Ozone Characteristics.* Springfield, VA: NTIS PB82-234212.

Koch, R. C. et al. (1977). *A Quality Modeling Study to Analyze the Impact of Various SO2 Control Strategies on Ambient Air Quality in the San Francisco Bay Area.* Springfield, VA: NTIS PB-274472/0GA.

Larsen, R. I. (1974). "A Mathematical Model for Relating Air Quality Measurements to Air Quality Standards." *Office of Air Programs Publications No. AP-89.* Office of Technical Information and Publications, U.S. Environmental Protection Agency. Springfield, VA: NTIS PB-205277.

Liu, Mei-Kao and Dale R. Durran (1977). *The Development of a Regional Air Pollution Model and Its Application to the Northern Great Plains.* Springfield, VA: NTIS PB-285980/9GA.

Mathis, J. J. Jr. et al. (1973). *A Review of Methods for Predicting Air Pollution Dispersion.* Springfield, VA: NTIS N73-20658.

McMaster, Larry R. (1982). *Air Quality Data Handling System (AQDHS-II). Test Run Series Documentation, 2nd Edition, Final Report.* Springfield, VA: NTIS PB82-256181.

Miller, C. W. et al (1973). *Validation of Annual Average Air Concentration Predictions from the AIRDOS-EPA Computer Code.* Springfield, VA: NTIS CONF-810462-2.

Miller, Catherine G. (1981). *Case Studies in the Application of Air Quality Modeling in Environmental Decision Making: Summary and Recommendations.* Springfield, VA: NTIS PB81-213233.

Mills, Michael T. et al. (1981). *Evaluation of Point Source Dispersion Models.* Teknekron Research Inc. for USEPA. Springfield, VA: NTIS PB82-121062.

NATO Committee on the Challenges of Modern Society (1980). *Practical Demonstration of Urban Air Quality Simulation Models.* Springfield, VA: NTIS PB82-233024.

Shreffler, J. H. (1982). *Observations and Modelling of NOx in an Urban Area.* Springfield, VA: NTIS PB83-140749.

Singpurwalla, Nozer D. (1975). "Models in Air Pollution," in *A Guide to Models in Governmental Planning and Operation,* ed. by Saul I. Gass and Roger L. Sisson. Potomac, MD: Sauger Books.

Trijonis, John, et al. (1982). *Validation of the EKMA Model Using Historical Air Quality Data, Final Report.* Springfield, VA: NTIS PB82-197187.

TRW Inc. (1972). *Prediction of the Effects of Transportation Controls on Air Quality in Major Metropolitan Areas.* Springfield, VA: PB-214176.

U.S. Air Force, Air Weather Service (1971). *Guide to Local Diffusion of Air Pollutants.* Technical Report 214. Springfield, VA: NTIS AD-726984.

U.S. Environmental Protection Agency (1982). *AEROS Manual Series.* Vol. I: *AEROS Overview Update No. 2.* Springfield, VA: NTIS PB82-240789.

U.S. Environmental Protection Agency (1976). *Compilation of Air Pollutant Emission Factors,* 2nd Ed. 2 Vols. Springfield, VA: NTIS PB-264194, PB-264195.

U.S. Environmental Protection Agency (1983). *ENAMAP-1A Long Term Air Pollution Model: Refinement of Transformation and Deposition Mechanisms.* Springfield, VA: NTIS PB83-140731.

U.S. Environmental Protection Agency (1983). *LONG2/SHORT2 Air Diffusion Models.* Springfield, VA: NTIS PB83-146118.

U.S. Environmental Protection Agency (1983). *Modifications to MOBILE2 Which Were Used by EPA to Respond to Congressional Inquiries on the Clean Air Act.* Springfield, VA: NTIS PB83-137414.

U.S. Environmental Protection Agency (1983). *1976 National Emissions Report: National Emissions Data System of the Aeronautic and Emissions Reporting System (AEROS).* Springfield, VA: NTIS PB83-114736.

U.S. Environmental Protection Agency (1983). *Observations and Modeling of NOx in an Urban Area.* Springfield, VA: NTIS PB83-130047.

U.S. Environmental Protection Agency (1981). *Regional Workshops on Air Quality Modeling: A Summary Report.* Springfield, VA: NTIS PB83-150573.

U.S. Environmental Protection Agency, Office of Research and Development (1979). *Sulfur Emission Control, Technology, and Waste Management.* Washington, DC: U.S. Environmental Protection Agency EPA-600/9-79-019.

User's Manuals

Benson, Paul E. (1979). *CALINE3—A Versatile Dispersion Model for Predicting Air Pollutant Levels Near Highways and Arterial Streets.* Springfield, VA: NTIS PB80-220841.

Bjorklund, Jay R. and James F. Bowers (1983). *User's Instructions for the SHORT2 and LONG2 Computer Programs,* Vol. I. Springfield, VA: NTIS PB83-146100.

Bjorklund, Jay R. and James F. Bowers (1983). *User's Instructions for the SHORT2 and LONG2 Computer Programs,* Vol. II. Springfield, VA: NTIS PB83-146092.

Burt, E. W. (1977). *Valley Model User's Guide.* Springfield, VA: NTIS PB-274054/ GGA.

Guthman, Lewis E. (1978). *User's Guide to MOBILE1 Mobile Source Emissions Model.* U.S. Environmental Protection Agency. Springfield, VA: NTIS PB81-159964.

Ingram, Gregory K. et al. (1974). *TASSIM: A Transportation and Air Shed Simulation Model.* Volume II: *Program User's Guide.* Springfield, VA: NTIS PB-232934.

Jones, K. E. et al. (1978). *A User's Manual for the CALINE-2 Computer Program.* Springfield, VA: PB-271106.

Kokin, Allan et al. (1973). *Controlled Evaluation of the Reactive Environmental Simulation Model (REM).* Volume II: *User's Guide.* Springfield, VA: NTIS PB-220457.

McMaster, Larry R. (1980). *Air Quality Data Handling System (AQDHS-II) Test Run Series Documentation.* Springfield, VA: NTIS PB82-256181.

Petersen, William B. (1978). *User's Guide for PAL: A Gaussian Plume Algorithm for Point, Area, and Line Sources.* Research Triangle Park, NC: U.S. Environmental Protection Agency, Office of Research and Development EPA-600/4-78-013.

Reifenstein, Edward C. et al. (1974). *Hackensack Meadowlands Air Pollution Study—AQUIP Software System User's Manual.* Environmental Research and Technology, Inc. Springfield, VA: NTIS PB-238605.

U.S. Environmental Protection Agency (1983). *CHAUG—A Program for Comparative Averages of Hourly Air Pollution Concentrations, User's Guide.* Springfield, VA: NTIS PB83-107342.

U.S. Environmental Protection Agency (1983). *PAL-DS Model: The PAL Model Including Deposition and Sedimentation User's Guide.* Springfield, VA: NTIS PB83-117739.

U.S. Environmental Protection Agency (1983). *MPTER-DS: The MPTER Model Including Deposition and Sedimentation User's Guide.* Springfield, VA: NTIS PB83-114739.

U.S. Environmental Protection Agency (1973). *User's Guide for the Climatological Dispersion Model.* Springfield, VA: NTIS PB-227346.

U.S. Environmental Protection Agency (1982). *Addendum to User's Guide for Climatological Dispersion Model.* Springfield, VA: NTIS PB82-246075.

U.S. Environmental Protection Agency (1975). *User's Guide for HIWAY, A Highway Air Pollution Model.* Springfield, VA: NTIS PB-239944/AS.

U.S. Environmental Protection Agency (1978). *User's Guide for PAL.* Research Triangle Park, NC: U.S. Environmental Protection Agency, EPA-600/4-78-013.

U.S. Environmental Protection Agency, Office of Air and Waste Management (1977). *User's Manual for Single-Source (CRSTER) Model.* Research Triangle Park, NC: U.S. Environmental Protection Agency EPA-450/2-77-013.

Willis, Byron H. (1973). *Hackensack Meadowlands Air Pollution Study. Evaluation and Ranking of Land Use Plans.* Springfield, VA: NTIS PB-238606.

LAND CAPABILITY EVALUATION

General Background

Dueker, Kenneth J. (1979). "Land Resource Information Systems: A Review of Fifteen Years of Experience," *Geo-Processing,* Vol. 1, p. 105–128.

Harvard University, Department of Landscape Architecture (1967). *Three Approaches to Environmental Resource Analysis.* Washington, DC: The Conservation Foundation.

Hills, G. Angus (1966). "The Classification and Evaluation of Land for Multiple Uses," *Forestry Chronicle,* June 1966, p. 1–25.

Lewis, Philip H. (1964). "Quality Corridors for Wisconsin," *Landscape Architecture Quarterly,* January 1964, p. 100–107.

Massachusetts Department of Community Affairs (1975). *Developing a Land Use Management Process.* Boston: Massachusetts Department of Community Affairs.

McHarg, Ian (1969). *Design with Nature.* New York: Natural History Press.

Schneider, Devon M. (1979). *Computer-Assisted Land Resources Planning.* Chicago: American Planning Association, Planning Advisory Service Report No. 339.

Sneath, Peter H. A. and Robert R. Sokal (1973). *Numerical Taxonomy.* New York: W. H. Freeman and Co.

Spangle, William and Associates, F. Beach Leighton and Associates, and Baxter McDonald and Company (1976). *Earth-Science Information in Land-Use Planning—Guidelines for Earth Scientists and Planners.* Washington, DC: U.S. Geological Circular 721.

Spangle, William and Associates (1974). *Application of Earth Science Information in Urban Land Use Planning.* Springfield, VA: NTIS PB-238081.

Models and Methods

Bishop A. B. et al. (1974). *Carrying Capacity in Regional Environmental Management.* Washington, DC: U.S. Environmental Protection Agency EPA 600/5-74-021.

Calkins, H. W. et al. (1977). *Geographic Information Systems, Methods and Equipment for Land Use Planning.* Springfield, VA: NTIS PB-286643.

Dobson, J. E. (1975). *Land Use Suitability Screening for Power Plant Sites in Maryland.* Springfield, VA: NTIS CONF-751122-1.

Fabos, Julius Gy., Christopher M. Greene, and Spencer A. Joyner Jr. (1978). *The METLAND Landscape Planning Process: Composite Landscape Assessment, Alternative Plan Formulation and Plan Evaluation.* Research Bulletin 643. Amherst, MA: University of Massachusetts, Massachusetts Agricultural Experimental Station.

Fromuth, Rick (1978). *Selection of Stormwater Models for Application to the Ohio Capability Analysis Program.* Columbus, OH: Ohio Department of Natural Resources.

Gordon, Steven I. (1978). "Performing land-capability evaluation by use of numerical taxonomy: land use and environmental decisionmaking made hard?", *Environment and Planning A,* Vol. 10, p. 915–921.

Gordon, Steven I. and Gaybrielle E. Gordon (1981). "The Accuracy of Soil Survey Information for Urban Land Use Planning," *APA Journal,* July, p. 301–312.

Hartigan, J. A. (1975). *Clustering Algorithms.* New York: John Wiley & Sons.

Interagency Council on Natural Resources and the Environment (1974). *Texas*

Natural Resources Information System Conceptual Design. Austin TX: Interagency Council on Natural Resources and the Environment.

Meyers, C. R. Jr., D. L. Wilson, and R. C. Durfee (1976). *Application of the Orrmis Geographical Digitizing and Information System Using Data from the Carats Project.* Springfield, VA: NTIS ORNL/RVS-12.

Miller, William R. (1975). *A Survey of Geographically Based Information Systems in California.* Sacramento, CA: Intergovernmental Board on Electronic Data Processing.

Nieswand, George H. and Peter J. Pizor (1977). *A Practical Carrying Capacity Approach to Land-Use Planning.* New Brunswick, NJ: Rutgers University.

Place, J. C. (1973). *Land Use Mapping and Modeling for the Phoenix Quadrangle.* Springfield, VA: NTIS N74-12129.

Putnam, Stephen H. (1976). *Laboratory Testing of Predictive Land Use Models Some Comparisons.* Springfield, VA: NTIS PB-265244.

Rogoff, Marc Jay (1978). *Statewide Computer Based Land Information Systems: An Annotated Bibliography of an Emergent Field.* Monticello, IL: Council of Planning Librarians, Exchange Bibliography 1490.

Rowe, J. Stan and John W. Sheard (1981). "Ecological Land Classification: A Survey Approach," *Environmental Management,* Vol. 5, No. 5, p. 451–464.

Rummel, R. J. (1970). *Applied Factor Analysis.* Evanston, IL: Northwestern University Press.

Sharky, Bruce G. and Richard C. Hall (1971). *A Study Program, Organization and Operation ERMS (Environmental Resource Management System).* San Diego, CA: San Diego County IREM Project (Integrated Regional Management).

U.S. Department of the Interior: Office of Land Use and Water Planning; U.S. Geological Survey (1976). *Critical Areas and Information/Data Handling.* New York: Carl Mays, Corkill & Seddon.

Urban Studies Center, University of Louisville (1971). *NEWCOM Summary.* Louisville, KY: Office of Economic Opportunity, Community Development Division.

User's Manuals

Cowen, David J., James N. Bayne and Daniel A. Fairey (1976). *Development and Applications of the South Carolina Computerized Land Use Information System.* Columbia, SC: Land Resources Conservation Commission.

Craig, Will (1976). *Minnesota Land Management Information System Geocoding Procedures 4005.* Minneapolis: University of Minnesota, Center for Urban and Regional Affairs and State Planning Agency.

Dueker, Kenneth et al. (1976). *Computer Maps and Geographic Data Analysis: An Exhibit Notebook.* Iowa City, IA: University of Iowa, Institute of Urban and Regional Research.

Earth Resources Data Analysis Systems (undated). *User's Manual.* 999 McMillan Street, Atlanta, GA 20218.

Edwards, R. G. and R. C. Durfee (1976). *Digitizing Geographic Data with GRIDOT, A Generalized Program for Drawing Overlay Grids in Various Map Projections.* Springfield, VA: NTIS ORNL/RUS-17.

Finch, Peter N., et al. (1982). *Interactive Computer-Aided Water Resources and Environmental Planning Using Color Raster Graphics: Program Description and User's Guide*. Springfield, VA: NTIS PB83-169250.

Frankland, Phillip (1976). *User's Manual for Computer Mapping Programs: POWRMAP and SIGNMAP*. Iowa City, IA: University of Iowa, Institute for Urban and Regional Research.

Gordon, Gaybrielle (1978). *User's Guide to the Ohio Capability Analysis Program*. Columbus, OH: Ohio Department of Natural Resources.

Hicks, Jimmy E. and Tom Hauger (1977). *Managing Natural Resource Data: Minnesota Land Management Information System*. Lexington, KY: The Council of State Governments.

Holmes, David D. and Rebecca L. Jolly (1980). *IMGRID version 3.5*. Norris, TN: Tennessee Valley Authority.

Mashburn, R. G., R. C. Durfee, and R. G. Edwards (1976). *REGRID: A Generalized Grid-to-Grid Transformation Procedure*. Springfield, VA: NTIS ORNL/RUS-18.

Nebraska Natural Resources Commission (1978). *Nebraska Natural Resources Data Bank Information System, Users Facilities Manual*. Lincoln, NE: Natural Resources Commission.

New York State Department of Commerce (1974). *LUNR Classification Manual*. Albany, NY: New York State Department of Commerce.

New York State Economic Development Board (1976). *User's Manual for PLAN-MAPII*. Albany: N.Y. State Economic Development Board.

Ohio Environmental Protection Agency (1980). *The PEMSO Nonpoint Source Screening Report No. 2*. Columbus, OH: Office of the Planning Coordinator, OEPA.

Ohio EPA (1979). *The PEMSO System Orientation Manual Report No. 1*. Office of the Planning Coordinator, Columbus, OH: Ohio EPA.

Stephenson, R. L. (1972). *CATCH: Computer Assisted Topography, Cartography, and Hypsography*. Part II: *ORGRAT: A General Subroutine for Drawing Graticules*. Springfield, VA: NTIS EDFB-1BP-72-7.

Stephenson, R. L. (1974). *SYMCON: A Program for Converting Latitude and Longitude to Orthogonal Coordinates*. Springfield, VA: NTIS ORNL-NSF-EP-76.

Stephenson, R. L. (1974). *TVMAP: A Program for Plotting Geographically Distributed Data*. Springfield, VA: NTIS ORNL-NSF-EP-75.

Veldman, Donald J. (1973). *Fortran Programming for the Behavioral Sciences*. New York: Holt, Rinehart and Winston.

Wilson, D. L. (1976). *CELNDX: A Computer Program to Compute Cell Indices*. Springfield, VA: NTIS ORNL-RUS-14.

Computer Program Acquisition

ERDAS is available from ERDAS Inc., Atlanta, GA. Hardware and software must be purchased together.

IMGRID and other related programs are available from the Harvard Laboratory for Computer Graphics and Spatial Analysis, 520 Gund Hall, 48 Quincy Avenue, Cambridge, MA 02138. Write for a price list and catalogue.

New York State Department of Commerce (1978). *LUNR Point and Area Data Overlays Ordering Folder.* New York Department of Commerce, 99 Washington Avenue, Albany, NY 12245. (List of publications and documentation can be ordered.)

OCAP is available under a licensing agreement. Contact Wayne Channel, OCAP, Soil and Water Conservation Districts Office, Ohio Department of Natural Resources, Columbus, OH 43215.

HAZARDOUS WASTE DISPOSAL

General Background

Alexander, George et al. (1977). *Toxic Substances Issue: Environment Midwest.* Chicago: Region V, U.S. Environmental Protection Agency.

Baski, Henry A. (1979). "Ground-Water Computer Models—Intellectual Toys," *Ground Water,* Vol. 17, No. 2 (March–April), p. 177–179.

Brown, Michael H. (1979). "Love Canal, U.S.A.," *New York Times Magazine,* January 21, p. 23.

Carter, Luther J. (1979). "An Industry Study of TSCA: How to Achieve Credibility?" *Science,* Vol. 203, January 19, p. 247–249.

Cherry, J. A., R. W. Gillham, and J. F. Pickens (1981), "Contaminant hydrogeology: 1, Physical processes," *Geoscience Canada,* Vol. 2, No. 2, p. 76–84.

Council for Agricultural Science and Technology (1976). *Application of Sewage Sludge to Cropland: Appraisal of Potential Hazards of the Heavy Metals.* Springfield, VA: NTIS PB-264015.

Fochtman, Edward G. et al. (1979). *Treatability of Carcinogenic and Other Hazardous Organic Compounds.* Cincinnati, OH: Municipal Environmental Research Laboratory, USEPA EPA-600/2-79-097.

Freeze, R. A. (1972). "Subsurface hydrology at waste disposal sites," *IBM Journal of Research and Development,* Vol. 16, No. 2, p. 117–129.

Fuhriman, Dean K. et al. (1971). *Ground Water Pollution in Arizona, California, Nevada, and Utah.* Springfield, VA: NTIS PB-211145.

Ghassemi, M. et al. (1976). *Disposal of Small Batches of Hazardous Wastes.* Cincinnati, OH: U.S. Environmental Protection Agency Report SW-562, 1976.

Gorelick, Stephen M. and Irwin Remson (1982). "Optimal Dynamic Management of Groundwater Pollutant Sources," *Water Resources Research,* Vol. 18, No. 1, p. 71–76.

Greenberg, Michael (1983). "Environmental Toxicology in the United States," in *Geographical Aspects of Health,* ed. by N. McGlasran and J. Dlunden. London: Academic Press. P. 157–174.

Hall, Clinton, W. et al. (1983). *Research for Groundwater Quality Management.* Springfield, VA: NTIS PB83-152256.

Kohan, Allen M. (1975). *A Summary of Hazardous Substance Classification*

Systems. Cincinnati, OH: U.S. Environmental Protection Agency, EPA/530/SW-171.

Lehman. John P. (1977). *Overview and Objectives of Hazardous Waste Management.* Washington, DC: U.S. Environmental Protection Agency.

Levin, James et al. (1977). *Assessment of Industrial Hazardous Waste Practices. Special Machinery Manufacturing Industries.* Springfield, VA: NTIS PB-265981.

Meyer, Charles F. (1973). *Polluted Groundwater: Some Causes Effects, Controls, and Monitoring.* Springfield, VA: NTIS PB-232117.

National Water Well Association (1971). *Proceedings of the National Ground Water Quality Symposium.* Springfield, VA: NTIS PB-214614.

Rosenberg, D. G. et al. (1976). *Assessment of Hazardous Waste Practice in the Petroleum Refining Industry* Springfield, VA: NTIS PB-259097.

Shultz, David (1980). *Disposal of Hazardous Waste.* Cincinnati, OH: Municipal Environmental Research Laboratory, U.S. Environmental Protection Agency EPA-600/9-80-101.

Straus, Matthew A. (1977). *Hazardous Waste Management Facilities in the United States-1977.* Cincinnati, OH: U.S. Environmental Protection Agency EPA/530/SW-146.3.

U.S. Department of Transportation (1978). *Emergency Action Guide for Selected Hazardous Materials.* Washington, DC: National Highway Traffic Safety Administration.

U.S. Environmental Protection Agency (1975). *Information about Hazardous Waste Management Facilities.* Cincinnati, OH: U.S. Environmental Protection Agency EPA/530/SW-145.

U.S. Environmental Protection Agency (1975). *Hazardous Waste Disposal Damage Reports.* Cincinnati, OH: U.S. Environmental Protection Agency EPA/530/SW-151.

U.S. Environmental Protection Agency (1975). *Hazardous Waste Disposal Damage Reports, Document No. 2.* Cincinnati, OH: U.S. Environmental Protection Agency EPA/530/SW-151.2.

U.S. Environmental Protection Agency (1976). *Pharmaceutical Industry Hazardous Waste Generation, Treatment and Disposal.* Washington, DC: U.S. Environmental Protection Agency.

U.S. Environmental Protection Agency (1983). *Physical Properties and Lead Testing of Solidified/Stabilized Industrial Wastes.* Springfield, VA: NTIS PB83-147983.

U.S. Environmental Protection Agency (1982). *Proceedings and Recommendations of the Workshop on Groundwater Problems in the Ohio River Basin, April 28–29, 1981.* Springfield, VA: NTIS PB82-153354.

U.S. Environmental Protection Agency, Region II (1977). *Regional Public Meetings on the Resource Conservation and Recovery Act of 1976.* U.S. Environmental Protection Agency, Region II EPA-SW-14P.

U.S. Environmental Protection Agency (1983). *Research for Groundwater Quality Management.* Springfield, VA: NTIS PB83-152256.

U.S. Environmental Protection Agency (1975). *State Program Implementation Guide: Hazardous Waste Surveys.* Cincinnati, OH: U.S. Environmental Protection Agency EPA/530/SW-160.

U.S. Environmental Protection Agency. *State Decision-Makers Guide for Hazardous Waste Management.* Washington, DC: U.S. Environmental Protection Agency Report SW-617.

U.S. Nuclear Regulatory Commission (1977). *Regulatory and Other Responsibilities as Related to Transportation Accidents.* Springfield, VA: NTIS PB-269625.

Walsh, John (1978). "EPA and Toxic Substances Law: Dealing with Uncertainty", *Science,* Vol. 202, November 10, p. 598–602.

Weinstein, Norman J. (1977). *Municipal-Scale Thermal Processing of Solid Wastes.* Springfield, VA: NTIS PB-263396.

Williams, R. et al. (1976). *Economic Assessment of Potential Hazardous Waste Control Guidelines for the Inorganic Chemical Industry.* Springfield, VA: NTIS PB-263210.

Models and Methods

Anderson, M. P. (1981). "Groundwater Quality Models—State of the Art". in *Proceedings and Recommendations of the Workshop on Ground Water Problems in the Ohio River Basin.* Cincinnati, OH: Ohio River Basin Commission.

Anderson, Mary P. (1979). "Using Models to Simulate the Movement of Contaminates Through Groundwater Flow Systems," *CRC Critical Reviews in Environmental Control,* Vol. 9, No. 2 (November), p. 97–153.

Appel, C. A. and J. D. Bredehoeft (1978). "State of ground-water modeling in the U.S. Geological Survey," *U.S. Geological Survey Report 737.* Washington, DC: U.S. Geological Survey.

Asbeck, E. L. and Y. Y. Haimes (1983). *Partitioned Multiobjective Risk Method (PMRMP).* Technical Report No. 83-7. Cleveland, OH: Center for Large Scale Systems and Policy Analysis, Case Western University.

Bachmat, Y., J. Bredehoeft, D. Holz, and S. Sebastian (1980). *Ground-water Management: The Use of Numerical Models.* Washington, DC: American Geophysical Union, Water Resources Monograph No. 5.

Barrett, Bruce R. (1978). "Controlling the Entrance of Toxic Pollutants into U.S. Waters," *Environmental Science and Technology,* Vol. 12, No. 2 (February), p. 154–162.

Bredehoeft, J. D. and G. F. Pinder (1970). "Digital analysis of areal flow in multiaquifer groundwater systems: A quasi three-dimensional model," *Water Resources Research,* Vol. 6, No. 3 (June), p. 883–888.

Callahan, Michael A. et al. (1979). *Water-Related Environmental Fate of 129 Priority Pollutants.* Washington, DC: U.S. Environmental Protection Agency EPA-440/4-79-029a.

Charbeneau, R. J. (1981). "Groundwater contamination transport with adsorption and ion exchange chemistry: Method of characteristics for the case without dispersion," *Water Resources Research,* Vol. 17. p. 705–713.

Darr, Russell E. (1979). "Ground-Water Computer Models—Practical Tools," *Ground Water,* Vol. 17, No. 2 (April), p. 174–176.

Davidson, J. M. et al. (1980). *Adsorption, Movement, and Biological Degradation of Large Concentrations of Selected Pesticides in Soils.* Cincinnati, OH: U.S. Environmental Protection Agency EPA-600/2-80-124.

Epler, J. L. et al. (1980). *Toxicity of Leachates.* Cincinnati, OH: Municipal Environmental Research Laboratory, U.S. Environmental Protection Agency, EPA-600/2-80-057.

Faust, Charles R. and James W. Mercer (1980). "Ground-Water Modeling: Numerical Models," *Ground Water,* Vol. 8, No. 4 (July–August), p. 395–409.

Faust, Charles R. et al. (1981). "Ground-Water Modeling: Recent Developments," *Ground Water,* Vol. 8, No. 6 (November–December), p. 569–577.

Fields, Timothy Jr. et al. (1975). *Landfill Disposal of Hazardous Wastes: A Review of Literature and Known Approaches.* Cincinnati, U.S. Environmental Protection Agency EPA/530/SW-165.

Gorelick, S. M. (1982). "A model for managing sources of groundwater pollution," *Water Resources Research,* Vol. 18, p. 772–781.

Gorelick, S. M. and L. Remson (1982). "Optimal dynamic management of groundwater pollutant sources," *Water Resources Research,* Vol. 18, p. 71–76.

Gorelick, S. M., L. Remson, and R. W. Cottle (1979). "Management model of a groundwater pollutant system with a transient pollution source," *Water Resources Research,* Vol. 15, p. 1243–1249.

Gorelick, Steven M. (1982). "A model for managing sources of groundwater pollution," *Water Resources Research,* Vol. 18, No. 4, p. 773–781.

Grisak, G. E. and J. F. Pickens (1980). "Solute transport through fractured media. 1: The effect of matrix diffusion," *Water Resources Research,* Vol. 16, p. 719–730.

Haimes, Y. Y. and W. A. Hall (1974). "Multiobjectives in water resources systems analysis: The surrogate worth trade-off method," *Water Resources Research,* Vol. 10, p. 615–624.

Hamilton, David A. (1982). *Groundwater Management Strategy for Michigan Groundwater Modeling: Selection, Testing and Use.* Vol. I. Springfield, VA: NTIS PB83-143206.

Hamilton, David A. (1982). *Groundwater Management Strategy for Michigan Groundwater Modeling: Selection, Testing, and Use.* Vol. II. Springfield, VA: NTIS PB83-143214, 1982.

Hatayama, H. K. et al. (1980). *A Method for Determining the Compatibility of Hazardous Wastes.* Cincinnati, OH: U.S. Environmental Protection Agency EPA-600/2-80-076.

Hiltz, Ralph H. et al. (1977). *Emergency Collection Systems for Spilled Hazardous Materials.* Springfield, VA: PB-272790.

Konikow, Leonard F. et al. (1974). "Modeling flow and chemical quality changes in an irrigated stream-aquifer system," *Water Resources Research,* Vol. 10, No. 3, p. 546–561.

Lewis, Barney D. et al. (1982). *Evaluative Ground-Water Solute-Transport Model at the Idaho National Engineering Laboratory.* Springfield, VA: NTIS PB82-204066.

Lorber, Matthew N. et al. (1982). *An Evaluation of Three Pesticide Runoff Loading Models.* Springfield, VA: NTIS PB83-130047.

Loucks, Daniel P. and Tery R. Stedinger (1983). *Interactive Modeling and Data*

Management for Predictive Surface and Groundwater Quality and Quantity. Springfield, VA: NTIS PB83-173221.

Lynch, C. J. and R. M. Kumar (1975). *A Critical Review of Six Hazard Assessment Models.* Springfield, VA: NTIS AD-A035599.

Maddock, Thomas (1981). *An Evaluation of the Effectiveness of Combining Economic and Physical Ground-Water Models.* Springfield, VA: NTIS PB82-236597.

Mercado, Abraham (1976). "Nitrate and chloride pollution of aquifers: A regional study with the aid of a single-cell model," *Water Resources Research,* Vol. 12, No. 4, p. 731–747.

Metcalf, Robert C. et al. (1971). "Model Ecosystem for the Evaluation of Pesticide Biodegradability and Ecological Magnification," *Environmental Science and Technology,* Vol. 5, No. 8 (August), p. 709–713.

Pickens, John F. and Gerald E. Grisak (1981). "Scale-dependent dispersion in a statified granular aquifer," *Water Resources Research,* Vol. 17, No. 4, p. 1191–1211.

Pickens, John F., et al. (1981). "Modeling of scale-dependent dispersion in hydrogeologic systems," *Water Resources Research,* Vol. 17, No. 6, p. 1701–1711.

Pinder, George F. and J. D. Bredehoeft (1968). "Application of the digital computer for aquifer evaluation," *Water Resources Research,* Vol. 4, No. 5, p. 1069–1093.

Prickett, Thomas A. (1979). "Ground-Water Computer Models—State of the Art," *Ground Water,* Vol. 17, No. 2 (April), p. 167–173.

Robinson G. D. et al. (1978). *Nature to Be Commanded: Earth-Science Maps Applied to Land and Water Management.* Washington, DC: U.S. Government Printing Office, 1978.

Smith, John H. et al. (1977). *Environmental Pathways of Selected Chemicals in Freshwater Systems.* Springfield, VA: NTIS PB-274548.

Stoner, K. M. (1982). *A Decision Model for Evaluating Land Disposal of Hazardous Wastes.* Springfield, VA: NTIS AD-A123742/9.

U.S. Environmental Protection Agency (1982). *An Evaluation of the Effectiveness of Combining Economic and Physical Ground-Water Models.* Springfield, VA: NTIS PB82-236597.

U.S. Environmental Protection Agency (1983). *Mathematical Model, SERATRA, for Sediment-Containment Transport in Rivers and Its Application to Pesticide Transport in Four Mile and Wolf Creeks in Iowa.* Springfield, VA: NTIS PB83-122671.

U.S. Environmental Protection Agency (1983). *A Screening Procedure for Assessing the Transport and Degradation of Solid Waste Constituents in Subsurface and Surface Waters.* Springfield, VA: NTIS PB83-117036.

Walton, William C. (1979). "Review of Leaky Artesian Aquifer Test Evaluation Methods," *Ground Water,* Vol. 17, No. 3 (June), p. 220–223.

Willis, Robert (1976). "Optimal groundwater quality management: well injection of waste waters", *Water Resources Research,* Vol. 12, No. 1, p. 47–53.

Willis, Robert (1976). "A planning model for the management of groundwater quality", *Water Resources Research,* Vol. 15, No. 6, p. 1305–1312.

Witherspoon, J. P. et al. (1976). *State-of-the-Art and Proposed Testing for*

Environmental Transport of Toxic Substances. Springfield, VA: NTIS ORNL/ EPA-1.

Wolfe, N. Lee et al. (1976). *Chemical and Photochemical Transformation of Selected Pesticides in Aquatic Systems.* Springfield, VA: NTIS PB-258846.

User's Manuals

Ball, J. W., E. A. Jenne, and D. K. Nordstrom (1979). "WATEQ2—a computerized chemical model for trace and major element speciation and mineral equilibria of natural waters," in *Chemical Modeling in Aqueous Systems,* ed. by E. A. Jenne. American Chemical Society Symposium Series #93, p. 815–835.

Konikow, L. F. and J. D. Bredehoeft (1978). "Computer Model of Two Dimensional Solute Transport and Dispersion in Ground Water," in *Techniques of Water-Resources Investigation of the United States Geological Survey,* Chapter C2, Book 7. Washington, DC: U.S. Government Printing Office.

Onishi, Y. et al. (1982). *User's Manual for the Instream Sediment-Contamination Transport Model SERATRA.* Springfield, VA: NTIS PB83-122739.

Perez, Armado I. et al. (1974). *A Water Quality Model for a Conjunctive Surface-Groundwater System.* Springfield, VA: NTIS EPA-600/5-74-013.

Prickett, T. A. and C. G. Lonquist (1971). "Selected Digital Computer Techniques for Groundwater Resource Evaluation," *Bulletin 55,* Illinois State Water Survey: Department of Registration and Education.

Townley, Lloyd R. and John L. Wilson (1980). *Description of and User's Manual for a Finite Element Aquifer Flow Model AQUIFEM-1.* Cambridge, MA: Ralph M. Parsons Laboratory for Water Resources and Hydrodynamics, Massachusetts Institute of Technology.

Trescott, P. C., et al. (1976). "Finite-Difference Model For Aquifer Simulation In Two Dimensions With Results of Numerical Experiments," in *Techniques of Water-Resource Investigations of the U.S. Geological Survey,* Book 7. Washington, DC: U.S. Government Printing Office.

U.S. Environmental Protection Agency (1983). *User's Manual for the Instream Sediment Containment Transport Model SERATRA.* Springfield, VA: NTIS PB83-122739.

U.S. Environmental Protection Agency (1983). *User's Manual for the Chemical Transport and Fate Model (TOXIWASP),* Version 1, by Robert B. Ambrose, Jr., Sam I. Hill, and Lee Mulkey. EPA-600/3-83-005. Athens, GA: Environmental Research Laboratory, USEPA.

U.S. Environmental Protection Agency (1982). *Water Quality Analysis Simulation Program (WASP) and Model Verification Program (MVP)—Documentation,* by D. N. DiToro, J. J. Fitzpatrick, and R. V. Thomann. Duluth, MN: Office of Research Laboratory, USEPA.

Wilson, John L., Lloyd R. Townley, and Antonio Sa da Costa (1979). *Mathematical Development and Verification of a Finite Element Flow Model AQUIFEM-1.* Cambridge, MA: Ralph M. Parsons Laboratory for Water Resources, Massachusetts Institute of Technology, 1979.

COMPUTER PROGRAM ACQUISITION

Townley, Lloyd R. and John L. Wilson. *Description of and User's Manual for A Finite Element Aquifer Flow Model AQUIFEM-1*. Cambridge, MA: Ralph M. Parsons Laboratory for Water Resources and Hydrodynamics, Massachusetts Institute of Technology, 1980. Price of code: $300.

U.S.G.S. Solute and Flow Models cited in Chapter 6 are available from U.S.G.S. WRD Attention Charles Showen, Mail Stop 437, National Center, U.S. Geological Survey, Reston, VA 22092. Approximate cost: $40 each.

Index

Cited in text.